DATE DUE

DEMCO 38-296

An Illustrated

Guide to

Iowa Prairie

Plants

 A BUR OAK ORIGINAL

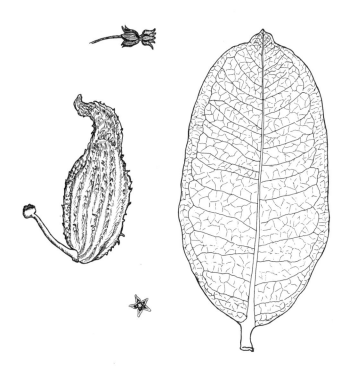

An Illustrated
Guide to
Iowa Prairie
Plants

Paul Christiansen

and

Mark Müller

University of Iowa Press

Iowa City

University of Iowa Press, Iowa City 52242

Copyright © 1999 by the University of Iowa Press

Printed in the United States of America

http://www.uiowa.edu/~uipress

Printed on acid-free paper

Library of Congress Cataloging-in-Publication Data

Christiansen, Paul.
    An illustrated guide to Iowa prairie plants / by
 Paul Christiansen and Mark Müller.
        p.   cm.—(A bur oak original)
    Includes bibliographical references and index.
    ISBN 0-87745-660-7, ISBN 0-87745-661-5 (pbk.)
    1. Prairie plants—Iowa—Identification.   I. Müller, Mark,
 1958–   . II. Title.   III. Title: Iowa prairie plants.   IV. Series.
 QK160.C48   1999
 581.7′44′09777—dc21   98-47433

99   00   01   02   03   C   5   4   3   2   1
99   00   01   02   03   P   5   4   3   2   1

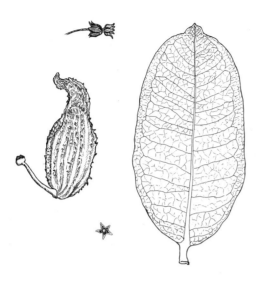

Contents

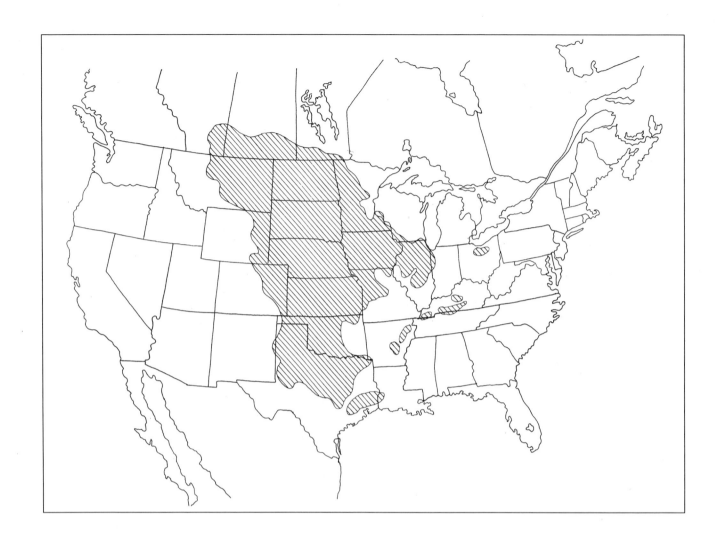

Preface and Acknowledgments

THE PRAIRIES OF IOWA and the states immediately adjacent to Iowa's borders are within the tallgrass prairie. The tallgrass prairie is on the eastern edge of the prairie formation or biome. This great belt of grassland stretches from the prairie provinces of Canada south to central Texas and west to the foothills of the Rocky Mountains.

Tallgrass prairie is found all along the eastern side of the prairie formation. The best large expanses still intact are found in the Kansas Flint Hills and south into eastern Oklahoma. The Sand Hills of central Nebraska also have large expanses of mixed prairie—tall, mid, and short grasses mixed depending upon moisture availability and exposure. Prior to European settlement, tallgrass prairie was concentrated in the "prairie peninsula" first described by Transeau in 1935.

The prairie peninsula extends from eastern Nebraska, southwestern Minnesota, and northwestern Missouri across Iowa and noses into the eastern forests of maple and oak across Illinois and southern Wisconsin, into Indiana, and even a bit into Ohio and Michigan. Iowa is the only state completely within the tallgrass prairie formation.

To the west of Iowa, across Kansas, Nebraska, and the Dakotas, the height of the vegetation becomes progressively shorter and is known as the midgrass prairie. Shortgrass prairies are found on the high plains and stretch across Montana, Wyoming, and Colorado toward the mountains.

The tallgrass prairie is very productive because of adequate summer rainfall and warm summer temperatures, although occasional severe drought occurs during the growing season. Grasses, sometimes with seed stalks over 6 feet tall, are common in late summer and fall. Sunflowers and compass plant nearly equal them in height. Hundreds of other species join them to produce a vegetation rich in diversity of color and form.

What should you expect to find when you visit a tallgrass prairie? Be ready for a large number of species at your feet. Be prepared for a tangle of grasses and forbs (flowers or nongrassy herbaceous plants) all around you. Expect to see something that resembles an unkempt hayland with little uniformity, except there are virtually no trees and hardly any shrubs. Get ready for many species to be in bloom anytime during the growing season, from early spring to frost and beyond in the fall. At any one time, most species will not be blooming, but many plants that bloomed prior to your visit will be producing fruits and setting seeds. Others will come into bloom later.

Grasses dominate the prairie. They furnish the matrix within which the other plants grow. Notice that the plants in flower are usually the tallest, overtopping those that are now producing fruits or those that are not yet ready to flower. Those waiting their turn will grow above those flowering when the time comes, making them better able to attract the pollinators or shed pollen to the winds. Be ready to lose yourself in contemplating the richness and variety of the prairie. And be ready to make frequent trips back to see the prairie dressed with other suits of clothes, all through the growing season from spring to fall.

Tallgrass prairie vegetation is dominated by three plant families: the grasses (Poaceae), the daisy or composite family (Asteraceae), and the legume or pea family (Fabaceae). Grasses sometimes predominate in the number of different species but always dominate in annual production. The bulk of the living material on the prairie is furnished by the grasses. Grasses have two basic growth forms. Some form sods by spreading by underground stems (rhizomes) from which upright shoots appear. Slough grass and bluejoint are excellent examples of this habit. Other grasses, particularly those that inhabit drier sites, are bunch grasses, where year-to-year spread is a fringe of growth at the margin of the plant. Prairie dropseed is a good example of this habit. Others like big bluestem and Indian grass are intermediate in sod formation. It is not uncommon to find clumps of big bluestem in a circle, which has resulted from the dying out of the center of a spreading clump while its perimeter continues to spread.

Grasses have very reduced flowers, no petals or other bright colors, and typically many flowers (called florets) in a compact to spreading tuft (inflorescence). They are wind-pollinated, shedding pollen from the stamens into the air; sticky hairy stigmas on the pistil catch the pollen, leading to fertilization and seed set.

Composites such as sunflower and aster are another important group. Typically, the largest number of species on a prairie will be from the composite family. These plants group their flowers into heads, often with only the marginal flowers of the head producing prominent petals. Forms range from the tall and robust saw-tooth sunflower to pussytoes only a few inches tall.

The third family, the legume or pea family, also has many species. Their characteristic sweetpea-type flowers, podlike fruits (seed cases), and compound leaves make them easy to identify. Their nitrogen-fixing capabilities are well known.

The remainder of the prairie flora is found in many other plant families; forty-four additional families are treated in this book. Some of the other important families are the milkweed, buttercup, rose, and figwort families. The next section lists all the species in this book by family.

Iowa's tallgrass prairie takes different forms in various landscapes. Most of Iowa was covered by upland prairie where moisture and soil conditions were favorable for plant growth. In these conditions big bluestem, Indian grass, and prairie dropseed do well. Where the land is low-lying the soil is moist more of the time, and more moisture-loving plants thrive. Bluejoint, slough grass, and saw-tooth sunflower do well here. On even wetter sites, where the prairie begins to grade into marsh, a sedge meadow often develops. Here sedges, blue flag, and swamp milkweed are characteristic. On hilltops and ridges where the soil is thinner, little bluestem and field goldenrod predominate. Sandy prairies have a distinct flora, with June grass, sand reed-grass, and hairy puccoon often being prominent. Dry, hilly prairies are found on both the east and west sides of Iowa. In the west the Loess Hills are a chain of hills formed from wind-blown silt stretching from north of Sioux City to below the Missouri border along the floodplain of the Missouri River. Little bluestem, side-oats grama, dalea, soapweed (yucca), and other species able to withstand drought do well here. In the hilly portion of northeastern Iowa so-called goat prairies are often found on exposed southwest-facing slopes. Again, little bluestem dominates, along with side-oats grama and aromatic aster.

Examples of tallgrass prairie are found across the state. The largest prairies are under protection by the state of Iowa, certain counties, and private organizations. Many small remnants exist on old railroad rights-of-way, pioneer cemeteries, abandoned country school grounds, and other waste places. A partial list of prairies open to the public is given at the end of the book.

Maintaining Iowa's prairies calls for an active role by their managers. Trees often invade prairies after periodic burning or mowing no longer takes place. Good examples are on the goat prairies of northeastern Iowa, where red cedar has overrun many small prairies. The western portion of the Loess Hills was devoid of trees prior to European settlement, according to early travelers' reports. Now, trees and brush have invaded nearly to the tops of some of the hills. Consequently, prairie preserves and remnants are usually managed with fire or mowing to combat woody invasion and favor prairie species. Burning is often done on a two- to five-year cycle, usually in the spring, although managers are trying fall and even summer burns as well. Efforts to leave refuges for animals with small populations, such as some butterflies, by burning small sections, burning only portions of critical habitats each year, and varying the season of burn all help to maintain as wide a diversity of species as possible.

Efforts to reconstruct tallgrass prairie have been going on in Iowa since the 1950s. Learning from range reseeding in the Great Plains during the 1930s, from the experiments at the University of Wisconsin Arboretum, and from other sources, prairie enthusiasts have had considerable success. The largest reconstruction attempted in Iowa began in 1993 at Walnut Creek (now Neal Smith) National Wildlife Refuge near Prairie City, Iowa, east of Des Moines. There the federal Fish and Wildlife Service has begun to revegetate up to 8,000 acres of agricultural land with prairie and savanna. In 1997 a prairie learning center was opened, and bison were introduced to the refuge. Many school districts, parks, and other agencies, as well as hundreds of private individuals and many corporations, have done the same on a smaller scale. John Deere Corporation in Cedar Falls and the Iowa Farm Bureau at its headquarters in Des Moines are two examples. At the University of Northern Iowa in Cedar Falls a prairie has been reconstructed, as well as at Cornell College, Grinnell College, Iowa State University, and others. Prairie reconstructions are beginning to be noticed on state and county roadways as well. The Iowa Department of Transportation (DOT) began evaluating use of prairie grasses on highway rights-of-way in the 1950s. A county roadside program originated at the University of Northern Iowa in 1988 aiding county roadside managers in controlling erosion and preventing weedy growth by using prairie species. The Iowa DOT has introduced prairie grasses and flowers on primary highways in many locations across the state. Several seed and plant producers in Iowa and surrounding states have turned to prairie species to furnish materials to a growing number of persons interested in using prairie as an alternative to maintaining large expanses of mowed lawn on their properties.

We seem to have come full circle in our interest in prairies. The farmers who settled Iowa in the 1800s were intent on replacing the prairie with cropfields and pastures. Only after prairies had almost disappeared did we realize that they offered more than academic interest. We found the prairie to be a place of beauty, of utility, of historical interest, and of regeneration for our psyche.

Prairie watching is addictive. The diversity on the landscape, from shoetop to horizon, holds one's attention to the exclusion of the passing of time and matters of pressing concern. Studying the prairie is satisfying, especially aesthetically, but also intellectually. Reading the landscape through the species present, observing the season of bloom, deciphering the structure of the flowers, recognizing the visiting pollinators, investigating the size and shape of seeds, and experiencing the thrill of germinating seedlings are all part of getting to know prairies.

I prepared the plant descriptions and distributions while on sabbatical leave from Cornell College as a fellow at University House, the University of Iowa. The distribution maps were prepared with a grant from the Living Roadway Trust Fund, Iowa Department of Transportation. The Living Roadway Trust Fund and the Iowa Science Foundation funded the preparation of illustrations.

I wish to thank Dr. Roger "Jake" Landers for introducing me to the tallgrass prairie and my wife, Barbara, and my sons, Dana and Scot, for persevering these many years with the prairie as a constant companion.

—*Paul Christiansen*

During the two years that I worked on *An Illustrated Guide to Iowa Prairie Plants*, my "office" consisted of a lawn chair and sketch pad. Traveling to nearly every prairie remnant left in the state, I intended to illustrate all the plants in their natural repose. In the end, several elusive species were drawn from herbarium specimens and photographs.

Paul and I were always amazed at the radical variability among individuals of a species as well as the similarities between certain species; this led to an expansion of the range of measurements and revisions of descriptions. The primary focus of the illustrations and descriptions is to provide anatomical detail to aid in field identification. But we hope that this book will also enhance the appreciation and care for an amazing and amazingly threatened ecosystem: the tallgrass prairie.

I would like to thank Daryl Smith and Larry Eilers for my prairie addiction, Diana Horton and Jennifer Bell of the University of Iowa Herbarium for their immense help, and Paul for inviting me along on this project. Finally, I would like to thank Valerie Cool, who offered much encouragement, patience, and great companionship on many memorable prairie forays.

—*Mark Müller*

An Illustrated

Guide to

Iowa Prairie

Plants

How to Use This Book

THIS BOOK IS DESIGNED to enable those interested in prairie plants in Iowa and surrounding states to make reasonably accurate identifications and to learn more about the distribution, structure, and natural history of plants on the prairie. The species are presented in order by family and within families by genus. Pteridophytes, the ferns and fern allies, are first, followed by the flowering plants, dicots first and then monocots.

PLANT NAMES

With some exceptions, common names and scientific names follow those used by Lawrence Eilers and Dean Roosa in *The Vascular Plants of Iowa*. Where another common name is frequently used, it is placed below the preferred common name. Also, if there is a scientific name that has been widely used in the past but is no longer preferred, it is placed below the preferred name. Common and scientific names of the plant families are also given.

PLANT DESCRIPTIONS

Species are described from the ground up: stem, leaf, inflorescence, flower, fruit, and habitat. As you compare plants in the field to the descriptions given, especially plant and leaf size, keep in mind that considerable variation occurs depending upon local conditions of competition, soil moisture, available nutrients, etc. Measurements in the descriptions were made on plants in the field and herbarium and give the average size range to be expected but are not meant to encompass every specimen encountered. The time of flowering and fruiting is given for the central part of Iowa. Differences of one or two weeks can be expected in the extreme northern and southern parts of Iowa for many species. Occasionally, additional information on the natural history of the species is given. Where several species are closely related, a common member of the group is fully described, and the other species are compared to the first in paragraph form with diagnostic characteristics italicized.

DISTRIBUTION MAPS

A distribution map for each species is included. Distribution maps are not intended to note the occurrence of every species in every county in which the species occurs. Rather they are to give a general idea of the range of each species across the state. These maps first appeared in a book printed and distributed in 1992 by the Iowa Department of Transportation, *Distribution Maps of Iowa Prairie Plants*, by Paul Christiansen.

DRAWINGS

Drawings of almost all species described or mentioned in the text are located close to the plant's description. In addition to the general shape of each plant, characteristic features of use in identification are included.

IDENTIFICATION

As an aid in identifying prairie plants, a guide to family identification appears in the section Family Finder. The aim is to speed identification by determining in which family a plant belongs, thus narrowing the number of drawings and descriptions one must scan to locate the correct species. By answering the questions and following the instructions, along with use of the species descriptions and illustrations, reliable identifications can usually be made. In order to help the reader become familiar with the plant families treated in this book, a short synopsis of each family is given following the Family Finder and on pages 186 and 200.

For more positive identifications plant manuals and keys should be consulted. Several that are useful for Iowa are Eilers and Roosa, *The Vascular Plants of Iowa*, which lists all verified species that exist in the state; Pohl, "Grasses of Iowa"; Van Bruggen, *The Vascular Flora of South Dakota*; Mohlenbrock, *Guide to the Vascular Flora of Illinois*; and Styermark, *Flora of Missouri*.

Where positive identification is important, taxonomists at local or state colleges or universities or other knowledgeable individuals should be consulted.

GLOSSARY

To make the descriptions as accurate as possible, some botanical terms have been used. To help in understanding the descriptions, a glossary of botanical terms precedes the Index.

SPECIES LIST

Below is a list of the species treated in this book. The order is pteridophytes (nonseed plants) followed by the angiosperms (flowering plants), with the dicots first followed by the monocots. The species are grouped by plant families in alphabetical order. Several families are now known by different names than in the past. The umbel family (Umbelliferae) is now the Apiaceae, the daisy family (Compositae) is the Asteraceae, the mustards (Cruciferae) are the Brassicaceae, the legumes (Leguminosae) are the Fabaceae, the mints (Labiatae) are the Lamiaceae, and the grasses (Gramineae) are the Poaceae. With some exceptions, all names follow Eilers and Roosa's *The Vascular Plants of Iowa.*

FAMILY	COMMON FAMILY NAME
SPECIES	COMMON NAME

Pteridophytes

Aspleniaceae	Spleenwort Family
Onoclea sensibilis	sensitive fern

Equisetaceae	Horsetail Family
Equisetum arvense	common horsetail
Equisetum laevigatum	smooth scouring-rush

Angiosperms: Dicotyledons

Apiaceae	Parsley Family
Cicuta maculata	water hemlock
Eryngium yuccifolium	rattlesnake master
Oxypolis rigidior	cowbane
Zizia aptera	heart-leaved meadow parsnip
Zizia aurea	golden alexanders

Apocynaceae	Dogbane Family
Apocynum androsaemifolium	spreading dogbane
Apocynum cannabinum	Indian hemp
Apocynum sibiricum	Indian hemp

Asclepiadaceae	Milkweed Family
Asclepias amplexicaulis	sand milkweed
Asclepias hirtella	tall green milkweed
Asclepias incarnata	swamp milkweed
Asclepias ovalifolia	oval milkweed
Asclepias sullivantii	prairie milkweed
Asclepias syriaca	common milkweed
Asclepias tuberosa	butterfly weed
Asclepias verticillata	whorled milkweed
Asclepias viridiflora	green milkweed

Asteraceae	Daisy Family
Achillea millefolium	western yarrow
Ambrosia psilostachya	western ragweed
Antennaria neglecta	pussytoes
Antennaria plantaginifolia	ladies'-tobacco
Artemisia campestris	western sagewort
Artemisia ludoviciana	white sage
Aster azureus	sky-blue aster
Aster ericoides	heath aster
Aster laevis	smooth blue aster
Aster lanceolatus	paniclad aster
Aster novae-angliae	New England aster
Aster oblongifolius	aromatic aster
Aster praealtus	willow aster
Aster sericeus	silky aster
Aster umbellatus	flat-topped aster
Brickellia eupatorioides	false boneset
Cacalia plantaginea	prairie Indian plantain
Cirsium altissimum	tall thistle
Cirsium discolor	field thistle
Cirsium flodmanii	Flodman's thistle
Cirsium hillii	Hill's thistle
Coreopsis palmata	tickseed
Echinacea angustifolia	
Echinacea pallida	pale coneflower
Echinacea purpurea	purple coneflower
Erigeron annuus	annual fleabane
Erigeron strigosus	fleabane
Euthamia graminifolia	lance-leaved goldenrod (with *Solidago*)
Helenium autumnale	sneezeweed
Helianthus grosseserratus	saw-tooth sunflower
Helianthus maximiliani	Maximillian's sunflower
Helianthus occidentalis	western sunflower
Helianthus rigidus	prairie sunflower
Helianthus tuberosus	Jerusalem artichoke
Heliopsis helianthoides	ox-eye
Heterotheca villosa	golden aster

Hieracium canadense	hawkweed	Campanulaceae	Harebell Family
Hieracium longipilum	hawkweed	*Campanula aparinoides*	marsh bellflower
Hieracium umbellatum	hawkweed	*Lobelia siphilitica*	great lobelia
Krigia biflora	false dandelion	*Lobelia spicata*	spiked lobelia
Lactuca canadensis	wild lettuce		
Lactuca ludoviciana	prairie lettuce	Caryophyllaceae	Pink Family
Lactuca tatarica	blue lettuce	*Silene antirrhina*	sleepy catchfly
Liatris aspera	rough blazing star		
Liatris cylindracea	cylindric blazing star	Cistaceae	Rockrose Family
Liatris ligulistylis	blazing star	*Helianthemum bicknellii*	frost weed
Liatris punctata	dotted blazing star		
Liatris pycnostachya	prairie blazing star	Euphorbiaceae	Spurge Family
Liatris squarrosa	scaly blazing star	*Euphorbia corollata*	flowering spurge
Lygodesmia juncea	rush-pink	*Euphorbia dentata*	toothed spurge
Machaeranthera spinulosa	cut-leaved goldenrod	*Euphorbia glyptosperma*	spurge
Nothocalais cuspidata	prairie dandelion	*Euphorbia marginata*	snow-on-the-mountain
Parthenium integrifolium	feverfew		
Prenanthes aspera	rough white lettuce	Fabaceae	Legume Family
Prenanthes racemosa	glaucous white lettuce	*Amorpha canescens*	lead plant
Ratibida columnifera	long-headed coneflower	*Amorpha nana*	fragrant false indigo
Ratibida pinnata	gray-headed coneflower	*Astragalus agrestis*	milk vetch
Rudbeckia hirta	black-eyed Susan	*Astragalus canadensis*	milk vetch
Rudbeckia subtomentosa	fragrant coneflower	*Astragalus crassicarpus*	ground plum
Senecio pauperculus	prairie ragwort	*Astragalus distortus*	bent milk vetch
Senecio plattensis	prairie ragwort	*Astragalus goniatus*	
Silphium integrifolium	rosinweed	*Astragalus lotiflorus*	milk vetch
Silphium laciniatum	compass plant	*Baptisia bracteata*	false indigo
Solidago canadensis	tall goldenrod	*Baptisia lactea*	white wild indigo
Solidago gigantea	smooth goldenrod	*Chamaecrista fasciculata*	partridge pea
Solidago missouriensis	Missouri goldenrod	*Crotalaria sagittalis*	rattle box
Solidago nemoralis	field goldenrod	*Dalea candida*	white prairie clover
Solidago riddellii	Riddell's goldenrod	*Dalea enneandra*	dalea
Solidago rigida	stiff goldenrod	*Dalea leporina*	foxtail dalea
Solidago speciosa	showy goldenrod	*Dalea purpurea*	purple prairie clover
Vernonia baldwinii	Baldwin's ironweed	*Dalea villosa*	silky prairie clover
Vernonia fasciculata	ironweed	*Desmanthus illinoensis*	prairie mimosa
		Desmodium canadense	showy tick trefoil
Boraginaceae	Forget-me-not Family	*Desmodium illinoense*	Illinois tick trefoil
Lithospermum canescens	hoary puccoon	*Glycyrrhiza lepidota*	wild licorice
Lithospermum caroliniense	hairy puccoon	*Lathyrus palustris*	marsh vetchling
Lithospermum incisum	fringed puccoon	*Lathyrus venosus*	veiny pea
Onosmodium molle	false gromwell	*Lespedeza capitata*	round-headed bush clover
		Lespedeza cuneata	silky bush clover
Brassicaceae	Mustard Family	*Lespedeza leptostachya*	prairie bush clover
Cardamine bulbosa	spring cress	*Lespedeza stipulacea*	Korean lespedeza
		Lespedeza virginica	slender bush clover
Cactaceae	Cactus Family	*Oxytropis lambertii*	locoweed
Opuntia fragilis	little prickly pear	*Pediomelum argophyllum*	silvery scurf-pea
Opuntia humifusa	eastern prickly pear	*Pediomelum esculentum*	prairie turnip

Psoralidium batesii	scurfy pea	*Oenothera laciniata*	ragged evening primrose
Strophostyles helvula	wild bean	*Oenothera perennis*	sundrops
Strophostyles leiosperma	wild bean	*Oenothera pilosella*	prairie sundrops
Tephrosia virginiana	goat's-rue	*Oenothera rhombipetala*	sand primrose
Vicia americana	vetch	*Oenothera villosa*	gray evening primrose
Vicia sativa	common vetch		
Vicia villosa		Oxalidaceae	Wood Sorrel Family
		Oxalis violacea	violet wood sorrel
Gentianaceae	Gentian Family		
Gentiana alba	pale gentian	Plantaginaceae	Plantain Family
Gentiana andrewsii	bottle gentian	*Plantago patagonica*	plantain
Gentiana puberulenta	downy gentian		
Gentianopsis crinita	fringed gentian	Polemoniaceae	Phlox Family
		Phlox bifida	cleft phlox
Geraniaceae	Geranium Family	*Phlox maculata*	wild phlox
Geranium carolinianum	cranesbill	*Phlox pilosa*	prairie phlox
Geranium maculatum	wild geranium		
		Polygalaceae	Milkwort Family
Hypericaceae	St. John's Wort Family	*Polygala sanguinea*	field milkwort
Hypericum punctatum	Spotted St. John's wort	*Polygala senega*	seneca snakeroot
Hypericum sphaerocarpum	round-fruited	*Polygala verticillata*	whorled milkwort
	St. John's wort		
		Polygonaceae	Smartweed Family
Lamiaceae	Mint Family	*Polygonum pensylvanicum*	Pennsylvania smartweed
Lycopus americanus	water horehound		
Monarda fistulosa	wild bergamot	Primulaceae	Primrose Family
Monarda punctata	spotted horsemint	*Dodecatheon meadia*	shooting star
Prunella vulgaris	self heal	*Lysimachia ciliata*	fringed loosestrife
Pycnanthemum pilosum	hairy mountain mint		
Pycnanthemum tenuifolium	slender mountain mint	Ranunculaceae	Buttercup Family
Pycnanthemum virginianum	common mountain mint	*Anemone canadensis*	Canada anemone
Scutellaria parvula	skullcap	*Anemone cylindrica*	windflower
Teucrium canadense	germander	*Delphinium virescens*	prairie larkspur
		Pulsatilla patens	pasque flower
Linaceae	Flax Family	*Ranunculus fascicularis*	early buttercup
Linum rigidum	stiff flax	*Ranunculus pensylvanicus*	bristly crowfoot
Linum sulcatum	wild flax	*Thalictrum dasycarpum*	purple meadow-rue
Lythraceae	Loosestrife Family	Rhamnaceae	Buckthorn Family
Lythrum alatum	winged loosestrife	*Ceanothus americanus*	New Jersey tea
		Ceanothus herbaceus	redroot
Nyctaginaceae	Four-o'clock Family		
Mirabilis hirsuta	hairy four-o'clock	Rosaceae	Rose Family
Mirabilis nyctaginea	wild four-o'clock	*Fragaria virginiana*	wild strawberry
		Geum triflorum	prairie smoke
Onagraceae	Evening Primrose Family	*Potentilla arguta*	tall cinquefoil
Calylophus serrulatus	toothed evening primrose	*Rosa arkansana*	sunshine rose
Epilobium coloratum	cinnamon willowherb	*Rosa blanda*	meadow rose
Gaura biennis	biennial gaura	*Rosa carolina*	pasture rose
Gaura coccinea	scarlet gaura	*Spiraea alba*	meadowsweet

Rubiaceae — Bedstraw Family
  Galium boreale — northern bedstraw
  Galium obtusum — wild madder

Salicaceae — Willow Family
  Salix discolor — pussy willow
  Salix humilis — prairie willow
  Salix petiolaris — shrub willow

Santalaceae — Sandalwood Family
  Comandra umbellata — bastard toadflax

Saxifragaceae — Saxifrage Family
  Heuchera richardsonii — alumroot
  Saxifraga pensylvanica — swamp saxifrage

Scrophulariaceae — Figwort Family
  Castilleja coccinea — Indian paintbrush
  Castilleja sessiliflora — downy painted cup
  Pedicularis canadensis — lousewort
  Pedicularis lanceolata — swamp lousewort
  Penstemon digitalis — foxglove penstemon
  Penstemon grandiflorus — large-flowered beardtongue
  Penstemon pallidus — pale beardtongue
  Scrophularia lanceolata — figwort
  Veronicastrum virginicum — Culver's root

Solanaceae — Nightshade Family
  Physalis heterophylla — ground cherry
  Physalis virginiana — Virginia ground cherry

Urticaceae — Nettle Family
  Pilea fontana — bog clearweed

Verbenaceae — Vervain Family
  Verbena hastata — blue vervain
  Verbena simplex — narrow-leaved vervain
  Verbena stricta — hoary vervain
  Verbena urticifolia — white vervain

Violaceae — Violet Family
  Viola pedata — bird's-foot violet
  Viola pedatifida — prairie violet
  Viola pratincola — common blue violet

Angiosperms: Monocotyledons

Agavaceae — Yucca Family
  Yucca glauca — soapweed

Commelinaceae — Spiderwort Family
  Tradescantia bracteata — spiderwort
  Tradescantia ohiensis — Ohio spiderwort

Cyperaceae — Sedge Family
  Carex bicknellii — sedge
  Carex brevior — sedge
  Carex gravida — sedge
  Carex lasiocarpa — slender sedge
  Carex muhlenbergii — sedge

Iridaceae — Iris Family
  Iris shrevei — blue flag
  Sisyrinchium campestre — blue-eyed grass

Liliaceae — Lily Family
  Allium canadense — wild onion
  Allium cernuum — nodding wild onion
  Allium stellatum — wild prairie onion
  Hypoxis hirsuta — yellow stargrass
  Lilium michiganense — Michigan lily
  Lilium philadelphicum — wood lily
  Melanthium virginicum — bunch-flower
  Zigadenus elegans — white camass
  Zigadenus glaucus — white camass

Orchidaceae — Orchid Family
  Cypripedium candidum — small white lady's-slipper orchid
  Platanthera leucophaea — eastern prairie fringed orchid
  Platanthera praeclara — western prairie fringed orchid
  Spiranthes cernua — nodding ladies'-tresses

Poaceae — Grass Family
  Agropyron smithii — western wheatgrass
  Agropyron trachycaulum — slender wheatgrass
  Andropogon gerardii — big bluestem
  Bouteloua curtipendula — side-oats grama
  Bouteloua gracilis — blue grama
  Bouteloua hirsuta — hairy grama
  Bromus kalmii — Kalm's bromegrass

Calamagrostis canadensis	bluejoint
Calamovilfa longifolia	sand reed-grass
Dichanthelium acuminatum var. *implicatum*	rosette panic grass
Dichanthelium leibergii	Leiberg's panic grass
Dichanthelium linearifolium	rosette panic grass
Dichanthelium oligosanthes var. *scribnerianum*	Scribner's panic grass
Dichanthelium oligosanthes var. *wilcoxianum*	Wilcox's panic grass
Elymus canadensis	Canada wild rye
Festuca paradoxa	fescue
Koeleria macrantha	June grass
Muhlenbergia cuspidata	plains muhly
Muhlenbergia racemosa	marsh muhly
Panicum virgatum	switchgrass
Phalaris arundinacea	reed canary grass
Poa palustris	fowl meadow grass
Schizachyrium scoparium	little bluestem
Sorghastrum nutans	Indian grass
Spartina pectinata	slough grass
Sporobolus asper	dropseed
Sporobolus heterolepis	prairie dropseed
Stipa spartea	porcupine grass
Stipa viridula	green needlegrass

FAMILY FINDER

The order of presentation in this book is by plant families. Recognition of plant families will greatly aid in identifying unknown plants. Following is a guide to identifying plant families found in this book. Compare the unknown plant with the phrases on the left margin. Look for the indented phrases below the most correct one for the next set of phrases. Continue until the most correct phrase ends with a family name. Turn to the next section (or to pages 186 and 200 for the sedges and grasses) and compare the plant with the family description or compare the unknown plant with the drawings and descriptions of plants in that family. If the match is not correct, come back to the Family Finder and choose an alternate pathway. This finder is not foolproof or universal but only a guide to help in locating unknowns.

Plants without flowers and producing spores instead of seeds (**Pteridophytes**)
    with broad leaves from a short stem, spores borne on separate stalks in beadlike structures
        Spleenwort Family (Aspleniaceae)
    with tiny leaves ensheathing the stem at the nodes, stems green, spores borne on cones at the stem tips
        Horsetail Family (Equisetaceae)

Plants producing flowers and seeds (**Angiosperms**)
    leaves net-veined, flowers usually with four or five parts or multiples of four or five or many parts
        Dicots (Dicotyledoneae)
    leaves linear with parallel veins, flowers often with three or six parts or reduced or absent
        Monocots (Monocotyledoneae)
        following Dicots
Note: rattlesnake master in the parsley family (a dicot) also has linear leaves and tiny flowers in heads.

DICOTS

FLOWER CHARACTERISTICS

Flowers in heads
    with overlapping fillaries
              Daisy Family (Asteraceae)
    with compound leaves and irregular flowers
              Legume Family (Fabaceae)
    with opposite leaves and square stems
              Mint Family (Lamiaceae)
    with tiny flowers
        with four-parted corolla, inferior ovary, and long
        tapering leaves
              rattlesnake master (*Eryngium*) in the Parsley
              Family (Apiaceae)
        with overlapping closed petals and swollen nodes
              Smartweed Family (Polygonaceae)
        with new flowers at the tip and old flowers dropping
        from below
              Milkwort Family (Polygalaceae)
        with the many stamens extending far beyond the petals
              prairie mimosa (*Desmanthus*) in the Legume
              Family (Fabaceae)
Flowers with petals fused into tubelike corollas
    with inferior ovaries
        large to medium blue flowers
              Harebell Family (Campanulaceae)
        small white flowers
              Bedstraw Family (Rubiaceae)
    with two-lipped corollas
        with square stems and opposite leaves
              Mint Family (Lamiaceae)
        with mostly opposite leaves but round stems
              Figwort Family (Scrophulariaceae)
    with regular spreading corollas
        with coiled inflorescence and stony fruits
              Forget-me-not Family (Boraginaceae)
        with long narrow corolla tube and spreading petal-tips
              Phlox Family (Polemoniaceae)
        with downward-facing flowers
              Primrose Family (Primulaceae)
        with stamens close to the pistil
              Nightshade Family (Solanaceae)
        with stamens adhering to the corolla tube
              Vervain Family (Verbenaceae)

Flowers forming a basal floral tube and with separate
marginal petals
    with the tube flaring from the base
              Rose Family (Rosaceae)
    with the tube narrow to the top
              Saxifrage Family (Saxifragaceae)
Flowers with irregular (zygomorphic) corollas
    with pinnately compound leaves
              Legume Family (Fabaceae)
    with two-lipped flowers
        with square stems and opposite leaves
              Mint Family (Lamiaceae)
        with mostly opposite leaves but round stems
              Figwort Family (Scrophulariaceae)
    with the lower petal forming a sac at the base of
    the corolla
              Violet Family (Violaceae)
Flowers with regular (actinomorphic) corollas and
separate petals
    with many petals and stamens
              Cactus Family (Cactaceae)
    with five petals and many stamens
              Rose Family (Rosaceae)
    with yellow petals, brown stamens, and linear pods
              partridge pea (*Chamaecrista*) in the Legume
              Family (Fabaceae)
Flowers with inferior ovaries
    with flowers in heads
              Daisy Family (Asteraceae)
    with flowers in umbels
              Parsley Family (Apiaceae)
    with flowers with many stamens
        with many petals and spines on the ovary
              Cactus Family (Cactaceae)
        with a red round fruit (hip)
              rose (*Rosa*) in the Rose Family (Rosaceae)
    with four petals and a four-lobed stigma
              Evening Primrose Family (Onagraceae)
    with two-lobed ovary beneath small four-parted corolla
              Bedstraw Family (Rubiaceae)

Flowers that are small
  without petals or sepals
    shrubs with flowers in racemes
           Willow Family (Salicaceae)
    small herbaceous plants with transparent stems
           Nettle Family (Urticaceae)
    with greenish flowers on a long raceme and basal leaves
           Plantain Family (Plantaginaceae)
  with tiny flowers
    with four-parted corolla, inferior ovary, and long
    tapering leaves
           rattlesnake master (*Eryngium*) in the Parsley
           Family (Apiaceae)
    with overlapping closed "petals" and swollen nodes
           Smartweed Family (Polygonaceae)
    with new flowers at the tip and old flowers dropping
    from below
           Milkwort Family (Polygalaceae)
    with petals and sepals usually not developing, flowers
    clustered in the upper leaf axils
           frost weed (*Helianthemum*) in the Rockrose
           Family (Cistaceae)

PLANT CHARACTERISTICS

Plants with milky juice in leaves and stems
  with flowers in umbels
           Milkweed Family (Asclepiadaceae)
  with small drooping flowers in branching clusters
           Dogbane Family (Apocynaceae)
  with stalked pistilate flower and stamens grouped into
  pseudoflowers with marginal petal-like bracts or glands
           Spurge Family (Euphorbiaceae)
Plants with basal leaves and leafless flower stalks (or a few
greatly reduced stem leaves)
  with flowers in heads
           Daisy Family (Asteraceae)
  with ray flowers only around the margin of the head
           prairie ragwort (*Senecio*)
  with ray flowers throughout the head (and milky juice
  in the stems and leaves)
    with narrow linear leaves
           prairie dandelion (*Nothocalais*)
    with broad dandelion-like leaves
           false dandelion (*Krigia*)
  with flowers in a drooping umbel
           shooting star (*Dodecatheon*) in the Primrose
           Family (Primulaceae)
  with tiny flowers in a long slender raceme
           Plantain Family (Plantaginaceae)

Plants with woody stems and branches
  with flowers with sepals and petals
    with small irregular flowers grouped into tight racemes
           lead plant (*Amorpha*) in the Legume Family
           (Fabaceae)
    with small regular white flowers
      in loose clusters
           New Jersey tea (*Ceanothus*) in the Buckthorn
           Family (Rhamnaceae)
      in tight clusters
           meadowsweet (*Spiraea*) in the Rose Family
           (Rosaceae)
    with large pink (sometimes white) flowers
           rose (*Rosa*) in the Rose Family (Rosaceae)
  with flowers without petals and sepals and often
  with stipules
           Willow Family (Salicaceae)
Plants with opposite leaves
  with two-lipped corollas
    with square stems
           Mint Family (Lamiaceae)
    with round stems
           Figwort Family (Scrophulariaceae)
  with shiny leaves
           Gentian Family (Gentianaceae)
  with milky juice
    with flowers in umbels
           Milkweed Family (Asclepiadaceae)
    with small drooping flowers in branching clusters
           Dogbane Family (Apocynaceae)
    with stalked pistilate and staminate flowers grouped
    into pseudoflowers
           Spurge Family (Euphorbiaceae)
  with downward-facing flowers
           Primrose Family (Primulaceae)
  with narrow corolla tubes and spreading petal tips
           Phlox Family (Polemoniaceae)
  with flowers in heads with fillaries below
           Daisy Family (Asteraceae)
  with small yellow flowers with many stamens
           St. John's Wort Family (Hypericaceae)
  with black sticky regions on the stem below the flowers
           sleepy catchfly (*Silene*) in the Pink
           Family (Caryophyllaceae)
  with flowers with spreading bracts resembling sepals and
  with blue sepals resembling petals
           Four-o'clock Family (Nyctaginaceae)
  with flowers with five yellow petals on first flowers and
  without petals or sepals in later flower
           Rockrose Family (Cistaceae)

with large yellow tubular flowers with inferior ovary
 Evening Primrose Family (Onagraceae)
with blue flowers in tight racemes
 Vervain Family (Verbenaceae)
with delicate toothed leaves and small green flowers
 Nettle Family (Urticaceae)

MONOCOTS

Plants with grasslike leaves and florets within tiny leaflike bracts
 Stems round, leaves alternating in two rows (ranks) on opposite sides of the stem, one or more florets in each spikelet with two basal bracts (glumes)
 Grass Family (Poaceae)
 Stems often triangular, leaves in three rows (ranks) up the stem, spikelets each with a single floret, each having a basal bract; spikelets usually in round or elongate spikes
 Sedge Family (Cyperaceae)
Plants usually with wider leaves and producing ordinary flowers
 Flowers with six similar parts (three sepals and three petals)
 ovary of the flower inferior (below the petals and sepals)
 yellow stargrass (*Hypoxis*) in the Lily Family (Liliaceae)
 ovary of the flower superior (within the petals and sepals)
 plants with linear, sharp-tipped basal leaves and large white-petaled flowers
 Soapweed (*Yucca*) in the Yucca Family (Agavaceae)
 leaves of various lengths, flowers large or small and white or orange
 Lily Family (Liliaceae)
 Flowers with three sepals different from the three petals
 all sepals similar and all petals similar
 flowers large, with inferior ovary
 Iris Family (Iridaceae)
 flowers small with three petals inconspicuous sepals
 Spiderwort Family (Commelinaceae)
 not all petals or sepals similar
 Orchid Family (Orchidaceae)

FAMILY DESCRIPTIONS
PTERIDOPHYTES

Aspleniaceae: The spleenwort family produces leaves that are usually lobed or divided. The spores are produced in sporangia borne in clusters on the underside of the leaves. Sensitive fern produces spores on smaller, specialized leaves.

Equisetaceae: The horsetail family produces spores in cones at the tips of apparently leafless, ribbed, jointed stems. Tiny leaves ensheathe the stem at each node. Stems arise from rhizomes below the soil surface. In certain species some stems are without cones.

ANGIOSPERMS: DICOTYLEDONS

Apiaceae: The parsley or umbel family has flowers on stalks that all originate at the same point, forming an inflorescence called an umbel. Small four-petaled flowers are attached at the tops of two-parted inferior ovaries, which produce fruits with linear markings, wings, or bristles. Secondary umbels may be formed below the primary inflorescence stalk. Often the leaves are divided and have flaring bases sheathing the stem. Water hemlock and golden alexanders are typical umbels. Rattlesnake master is atypical in its linear, yuccalike leaves and tight-headed flowers.

Apocynaceae: The dogbane family is similar in many ways to the milkweeds, except that dogbanes have small flowers in a branching inflorescence. The fruits are long, narrow pods and the seeds are bearded. Milky juice is found in the stems and leaves; the leaves are opposite on the stems.

Asclepiadaceae: The milkweed family has distinctive flowers. The petals arch backward, but there is a growth, the hood, on the upper surface of each petal, and in most species a "horn" grows from each hood. The fruits or pods are also distinctive, as are the seeds—each has a tuft of hairs. All parts of the plant have abundant milky juice.

Asteraceae: The daisy or composite family contains plants whose flowers are grouped into heads with small bracts attached at the base of the head. The individual flowers are attached to a rounded or cone-shaped receptacle. Two types of flowers are present. Ray flowers have all five petals fused into a single strap-shaped structure, often with a toothed tip. Disk flowers have the petals fused into a cylinder with each petal evident as a small point at the top of the cylinder. The ovary is inferior, and the sepals are modified into hairs or points attached to the top of the single-seeded fruit. The flowers may be arranged in three ways in the heads. The group that includes wild lettuce (and dandelion) has only ray flowers in the heads; these plants also have milky juice in the leaves and stems. Most composites have ray flowers around the outer margin of the head with the center filled with disk flowers; sunflowers and asters typify this group. A small group that includes the blazing stars and thistles has only disk flowers in the heads.

Boraginaceae: The forget-me-not or borage family has small, regular flowers with five fused sepals and petals set individually in the axils of the upper leaves. Seeds are protected within bony fruit walls. The inflorescence is distinctive: it is coiled so the developing flower buds are protected, but as development progresses the stem uncoils to expose the flowers. The leaves alternate on the stem.

Brassicaceae: The mustard family's former name, Cruciferae, denotes the crosslike character of its flowers, which are made up of four petals, usually white or yellow. There are six stamens, two short and four long. The fruit is often an elongate pod with two rows of seeds. The leaves are often divided or lobed.

Cactaceae: The cactus family has large flowers with many petals and stamens. The sepals, petals, and stamens are attached atop a large ovary with spines on its surface. The stems are flattened and divided into segments in Iowa species, the prickly pears. The flowers are attached in a row along the upper margin of the upper stem segments.

Campanulaceae: The harebell or bellwort family has bell-shaped flowers with five sepals, five fused petals, and five stamens. *Lobelia* has more irregular petals. Harebells have inferior ovaries, dry fruits, and opposite leaves.

Caryophyllaceae: The pink family has five separate petals and sepals as well as five stamens. The pistil is superior and develops into a dry capsule. The leaves are opposite on the stems.

Cistaceae: The rockrose family has first flowers with five conspicuous petals and many stamens arranged in a raceme at the top of the plant. Later flowers are without obvious petals. The pistil is superior and produces a capsule surrounded by the calyx. Narrow alternate leaves are crowded on hairy stems.

Euphorbiaceae: The spurge family has cuplike structures that bear tiny male flowers consisting of a single stamen and a larger female flower with a single three-lobed pistil. The cup margin is modified into five petal-like lobes in flowering spurge, while other species have marginal glands. The leaves are alternate on the stem. All parts of the plant have milky juice.

Fabaceae: The legume or pea family has typical sweetpea-type flowers and pods as fruits. The pods split into two halves with a row of seeds attached along one seam. Often the leaves are pinnately divided; examples are false indigo and showy tick trefoil. Two smaller groups have more regular flowers in which all the petals are similar, but the fruits have the typical family characteristics; examples are partridge pea and prairie mimosa.

Gentianaceae: The gentian family has late-blooming plants that have tubular corollas with five petal points and five alternating points. The five stamens are attached to the inside of the corolla tube. The superior ovary enlarges and elongates as it matures within the drying corolla tube. The leaves are shiny and green, crowded, and opposite on the stem.

Geraniaceae: The geranium family has prominent flowers with five sepals and petals and ten stamens. The superior pistil elongates in fruit into a beak that discharges the seeds explosively. The leaves are palmately lobed and mostly basal.

Hypericaceae: The St. John's wort family has flowers with five sepals and petals, many stamens, and a superior pistil that matures into a capsule. The inflorescence is corymbiform. The leaves are opposite, sometimes with tiny oil glands (black or translucent dots).

Lamiaceae: The mint family (also known as Labiatae) is distinguished by opposite leaves, square stems, and two-lipped flowers. The flowers have two upper petals and three lower ones. There are two or four stamens, and the ovary is four-lobed, with the style originating between the lobes. The ovary matures into four nutlets.

Linaceae: The flax family has flowers with small spreading corollas with five fused petals and five stamens. The globular ovary expands into a spherical fruit that breaks up into five parts containing five to ten shiny, slippery, brown seeds.

Lythraceae: The loosestrife family has flowers with six sepals and six petal tips at the top of a floral tube and with twice as many stamens as petals. The superior ovary develops into a dry fruit within the floral tube. Small flowers are borne in the leaf axils of small leaves at the top of the plant. Oval leaves are alternate on the stem.

Nyctaginaceae: The four-o'clock family has clusters of tiny flowers with five fused sepals, no petals, five stamens, and a superior pistil producing a one-seeded fruit. Each cluster of flowers is within colored bracts that resemble sepals. The leaves are opposite on the stem.

Onagraceae: The evening primrose family has four separate petals and four stigma segments in the shape of a cross. The petals are attached atop an elongated ovary that matures into a four-part dry fruit that begins splitting at the upper end to release the seeds.

Oxalidaceae: The wood sorrel family has vase-shaped flowers with five petal tips and ten stamens of two different lengths. Five stigma branches radiate from the tip of the ovary. The fruit is a capsule that splits into five segments. The leaves are basal with three radiating leaflets.

Plantaginaceae: The plantain family has tiny flowers with four sepals, four petals, and four stamens crowded onto a leafless raceme. The superior pistil matures into a capsule within the calyx. The linear leaves are basal with parallel veins.

Polemoniaceae: The phlox family has its five petals fused into a corolla tube with five points. The five stamens adhere to and are hidden within the corolla tube. The ovary, located at the bottom of the corolla tube, is topped by an elongated style that extends beyond the corolla tube. The linear, pointed leaves are opposite each other on the stem.

Polygalaceae: The milkwort family has small, irregular flowers with five petal-like sepals and three petals. The inflorescence, a raceme, develops new flowers at the tip while flowers with mature seeds fall from the base of the inflorescence. The leaves are alternate on the stem.

Polygonaceae: The smartweed or knotweed family has small flowers clustered into a tight to loose elongated inflorescence. The colored or white sepals (there are no petals) closely overlap and the flower remains essentially closed. Tiny black seeds develop within the fruits. The leaves are distinctive in having a basal sheath of tissue (ocrea) that encircles the stem, which is swollen at each node (hence the name knotweed). Often the upper margin of the ocrea is useful in identifying the various species.

Primulaceae: The primrose family is distinguished by having its flowers face downward. The flowers have five sepals, five petals fused near the base, and five stamens. The ovary is superior and produces a capsule. The leaves are basal or opposite on the stem.

Ranunculaceae: Buttercup family flowers are regular and usually not elaborate except for prairie larkspur. The petals (showy sepals in *Anemone*) are separate; there are many stamens; and the numerous fruits are separate and one-seeded except, again, for prairie larkspur, in which the three fruits with numerous seeds are fused toward the base. The leaves are usually divided, often with multiple divisions, into three similar parts.

Rhamnaceae: The buckthorn family has tubular flowers with five sepal and petal tips and five stamens; a superior ovary develops three seeds. The flowers are arranged in a panicle producing dense clusters. The plants are woody with leaves alternate on the stem.

Rosaceae: The rose family has flowers with five separate petals and sepals, many stamens, and five to many pistils. The sepals, petals, and stamens are attached to the edge of a floral cup within which the pistils are located. In wild strawberry and tail cinquefoil, the numerous pistils are arranged in the center of the flower. *Rosa* is different in having an inferior ovary with the individual styles producing a dense brush in the center of the flower. The leaves are usually compound.

Rubiaceae: The bedstraw or madder family has small flowers with four fused petals attached atop the ovary, which has two parts. The ovary matures into a small, two-lobed fruit that is often bristly. The leaves are in whorls of four to six at the nodes.

Salicaceae: The willow family is made up of shrubs and trees with flowers reduced to stamens or pistil and a few hairs as a corolla. The flowers are crowded into catkins, which are often pendulous. Ripe fruits burst open and shed tiny seeds with cottony hairs. Stems are woody and leaves are alternate, often with prominent stipules. Plants are male or female.

Santalaceae: The sandalwood family has small, five-petaled flowers with adhering stamens forming a tube above the ovary. The ovary matures into a spherical capsule. The leaves are simple and alternate on the stem.

Saxifragaceae: The saxifrage family has five small petals arising at the margin of a floral tube that surrounds the ovary. Sepals and stamens also arise from the margin of the floral tube. In prairie species, the leaves are basal and the inflorescence is carried well above the leaves on a long stalk.

Scrophulariaceae: The figwort family has four or five petals fused into a two-lipped tube. There are two or four stamens; sometimes a fifth, sterile stamen is at the back of the flower, giving rise to the name beard tongue as in the pentemons. In some species the irregularity of the flower is not as pronounced and two petals differ only slightly from the other three; Culver's root is in this category. The fruit is a capsule; the leaves are alternate, opposite, or whorled on the stem.

Solanaceae: The nightshade family has flowers with five fused petals producing a funnel-shaped corolla. The superior ovary develops into a berry (a miniature tomato); in the case of the ground cherries the calyx enlarges along with the fruit to encase it. The leaves are alternate on the stem.

Urticaceae: The nettle family has tiny, four-lobed, greenish white flowers crowded on short, branched flower stalks arising from the leaf axils. The fruits are small and flat, surrounded by the calyx. The leaves are opposite with toothed margins.

Verbenaceae: The vervain family has petals fused into a corolla tube that flares outward toward the top. The four stamens are contained within the corolla tube. The ovary is four-parted; the style projects from the tip, producing four nutlets at maturity. The leaves are simple and opposite on the stem.

Violaceae: The violet family has irregular flowers with five petals. The lower petal bends back upon the flower stalk, forming a spur within which nectar is produced. Five sepals alternate with the petals, and five stamens develop within the corolla around the three-parted ovary. The superior ovary develops into a capsule. The basal leaves are heart-shaped or divided into a bird-foot pattern.

ANGIOSPERMS: MONOCOTYLEDONS

Agavaceae: The yucca family has large flowers on a stalk carried well above the leaves. The three sepals and petals are white; there are six stamens and a superior three-parted ovary. The leaves are basal, linear, firm, and sharp-tipped.

Commelinaceae: The spiderwort family has flowers with three inconspicuous sepals and three petals or two large petals and one that is smaller and greenish white. The flowers are borne on an inflorescence ensheathed by two bracts. The leaves are oval to linear.

Iridaceae: The iris family has petal-like sepals and petals of different shapes and with an inferior ovary. The fruits break into three linear, triangular parts with seeds in two stacks within each portion. The leaves are linear.

Liliaceae: The lily family has flowers with three sepals and three petals of similar color and size. There are six stamens. The pistil is within the corolla and breaks into three parts upon maturity. The leaves are linear and narrow.

Orchidaceae: The orchid family has very highly specialized flowers. The three sepals alternate with the more modified petals, especially the lower petal, which has been modified into an insect-landing platform and often contains a nectary in a spurlike growth. The fruits dry and split into three parts; the seeds are extremely tiny and numerous. The leaves are oval to linear and often basal.

Species Descriptions and Illustrations

PTERIDOPHYTES

ASPLENIACEAE
Spleenwort Family

Sensitive fern
Onoclea sensibilis L.

Stem: perennial; underground stem.

Sterile fronds: stalk 8″ long; blade triangular, 9″ by 9″, pinnately lobed, lobing nearly to the midvein on the lower half of the blade, less deeply lobed above; smooth above and below; margins of the lobes wavy; margins turning black in late summer.

Fertile fronds: stalk 8″; blade with several pinnately arranged lobes, appressed to the midvein; contracted lobes bearing spherical clusters of spore-producing structures (sporangia); light brown when developing, dark brown when mature; the plants often not producing fertile fronds; development begins as early as mid-May and continues until early July.

Habitat: common on wet prairies and marshes; also in wet, open places and in moist, open woods; usually found in patches.

EQUISETACEAE
Horsetail Family

Common horsetail
Equisetum arvense L.

Sterile stem: perennial; from an underground stem; 6″ to 2′ tall; jointed; highly branched with whorls of branches at the stem nodes; branches 3″ long, jointed; stems are green with a rough surface and with tiny ribs; sheath of tiny leaves encircling each node, with 1/16″ points.

Fertile stem: from an underground stem; 6″ to 8″ tall; without branches; whitish; appearing before the sterile stems in the spring and withering within a few weeks; sheath of leaves at each node with 1/2″ tips.

Cone: at the tip of the fertile stem; 1″ long by 3/16″ in diameter, pointed toward the tip, drooping at maturity, light brown; producing spores from tiny capsules (sporangia) underneath the shieldlike plates of the cone; spore production is completed before mid-May.

Habitat: common in moist to dry, open places, often with some disturbance, such as roadsides or railroad embankments; frequent on prairies; usually growing in patches.

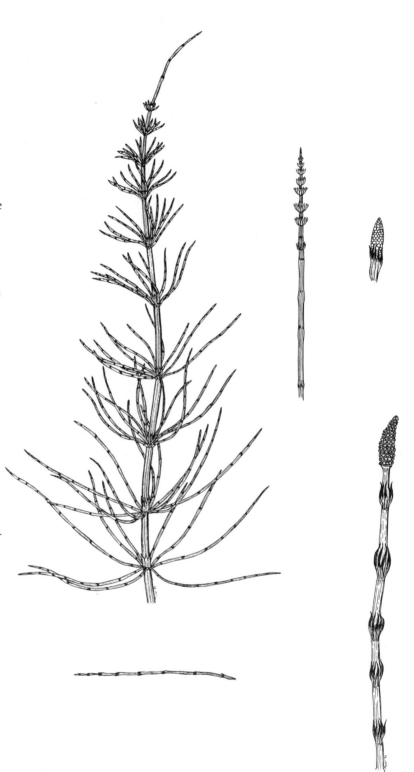

Smooth scouring-rush
Prairie scouring-rush
Equisetum laevigatum A. Br.

Stem: perennial; from an underground stem; 2′ to 4′ tall; jointed; unbranched; surface ribbed and very rough; sheath of tiny leaves at each node, with 1/16″ points, the points soon dropping; one dark band encircles each leaf sheath just below the points (at the top of the leaf sheath after the points drop).

Cone: at the tip of the stem; 3/4″ long by 3/16″ in diameter; cylindrical, tapering to an abrupt, pointed tip; producing spores from tiny capsules (sporangia) underneath the shield-like plates of the cone; cone development begins in late May and continues until late June; a few plants produce cones in the fall.

Habitat: common on mesic prairies; also on dry prairies and roadsides; usually growing in patches.

Another scouring-rush, *Equisetum hyemale* L., closely resembles smooth scouring-rush, but has two dark bands on each leaf sheath. It is found in more moist habitats and more often under trees.

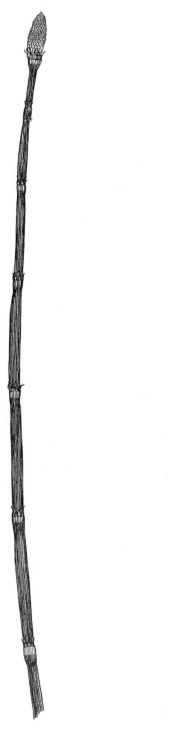

ANGIOSPERMS: DICOTYLEDONS

APIACEAE
Parsley Family

Water hemlock
Spotted water hemlock
Cicuta maculata L.

Stem: perennial; 3′ to 5′ tall; branching above; streaked with purple spots; smooth.

Leaves: alternate; twice compound; blade 8″ by 10″; leaf stalk about 6″ long, clasping the stem; leaflets with toothed margins; smooth above and below.

Inflorescence: compound umbels on flower stalks from the upper leaf axils; about 20 rays per umbel, about 20 flowers per ray; umbels to 3″ across.

Flowers: petals white, 1/16″ long; sepals 1/32″ long; petals and sepals attached to the upper part of double ovary; flowering from early to late July.

Fruits: two one-seeded fruits, 3/32″ long, prominent oil tubes on the fruit wall; fruiting begins in mid-July.

Habitat: frequent on wet to moist prairies.

All parts of this plant are very poisonous if ingested.

Rattlesnake master
Eryngium yuccifolium Michx.

Stem: perennial; 2′ to 3′ tall; unbranched; ridged, not hairy.

Leaves: alternate; linear, with clasping base and pointed tip; 4″ to 12″ by 1/2″ to 1″; prickles on the margins; smooth above and below.

Inflorescence: umbels of heads on flower stalks from the stem tip and upper leaf axils; heads 1/2″ tall by 1/2″ in diameter; spiny bracts below each flower in the head.

Flowers: petals white, 3/32″ long, inconspicuous; styles very long, white; flowering from mid-July to mid-August.

Fruits: two one-seeded fruits, 3/16″ long; with prominent oil tubes in the fruit wall; fruiting begins in early August.

Habitat: infrequent on moist to mesic prairies; a good indicator of the prairie plant community.

Cowbane

Oxypolis rigidior (L.) Raf.

Stem: perennial; 3′ to 5′ tall; unbranched, smooth.

Leaves: alternate; pinnately compound, seven to eleven leaflets; blade 3″ to 8″ long; leaf stalk 1″ to 3″, clasping the stem; leaflets 3″ by 1/2″ and smaller; margins with few, small teeth; smooth above and below.

Inflorescence: compound umbels on flower stalks from the stem tip and upper leaf axils; umbel to 4″ across; the flowers not crowded.

Flowers: petals white, 1/16″ long, attached to the top of the ovary; calyx tiny; flowering from early to late August.

Fruits: two one-seeded fruits, 3/16″ long; with wings from top to bottom; fruiting begins in mid-August.

Habitat: infrequent on moist prairies.

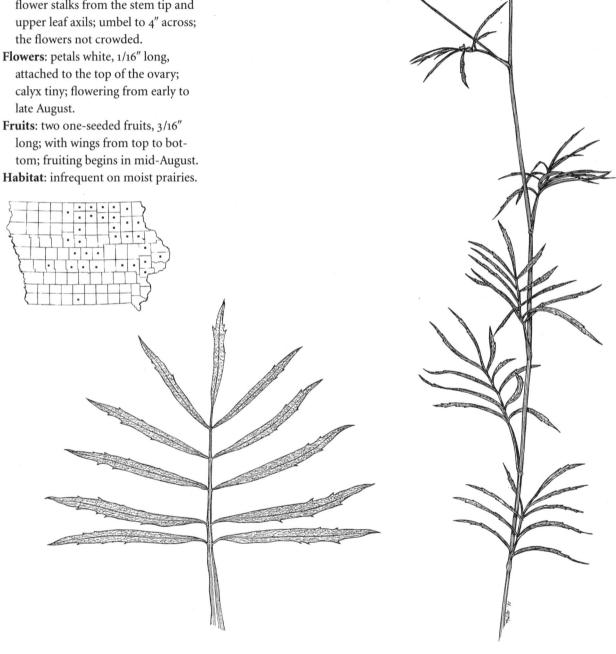

Golden alexanders
Zizia aurea (L.) W. Koch.

Stem: perennial; 1 1/2′ to 3′ tall; branched above; often several stems from the same root crown; smooth.

Leaves: basal and alternate; twice divided into threes; blade 3″ to 5″ long, end leaflet 1″ by 1/2″; basal leaves with long (8″) leaf stalks, stem leaves sessile; margins fine-toothed; smooth above and below.

Inflorescence: compound umbels on flower stalks from the stem tip and upper leaf axils; center flower in each secondary umbel sessile; flowers clustered but becoming more separated in fruit.

Flowers: petals yellow, 1/32″ long, set on a tiny ovary; sepals tiny; flowering from mid-May to mid-June.

Fruits: pairs of one-seeded fruits, 5/32″ long, flattened on one side, light brown, prominent oil tubes in the fruit wall; fruiting begins in early June; fruits mature and begin to drop in mid-August.

Habitat: common on moist to mesic prairies, in open woods and woodland openings; also on roadsides and in other open places.

Heart-leaved meadow parsnip
Zizia aptera (Gray) Fern.

Zizia aptera is similar to *Z. aurea* except the *lowest leaves are undivided and heart-shaped*. The leaf blades are 2″ by 1 1/2″, and the leaf stalk is about 6″ long. The inflorescence, flowers, and fruits are similar. Flowering is from mid-May to mid-June. Fruiting begins in early June. *Z. aptera* is infrequent on moist prairies.

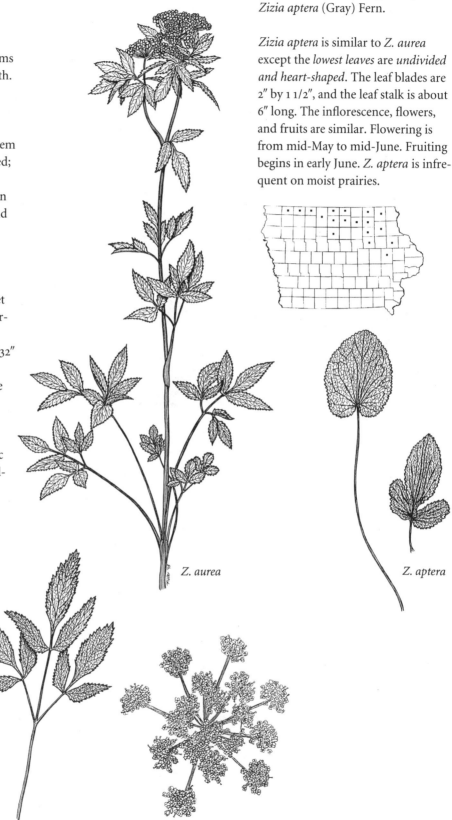

Z. aurea

Z. aptera

Z. aurea

APOCYNACEAE
Dogbane Family

Indian hemp
Apocynum sibiricum Jacq.

Stem: perennial; 2 1/2′ to 4′ tall; branched near the top; reddish; milky juice in the stem and leaves; smooth.

Leaves: opposite; squarish leaf shape with parallel sides; heart-shaped to rounded bases and abrupt, sharp tips; 2 1/2″ to 3″ by 1 1/8″; leaf stalks very short to absent; smooth above and below.

Inflorescence: small clusters of flowers on branching flower stalks from the stem tip and upper leaf axils.

Flowers: corolla white, 1/8″ long with five petal-tips; calyx 1/16″ long; flowering from late June to early August.

Fruits: seedpods are long, narrow follicles; 3″ long by 3/32″ in diameter; usually in pairs; seeds with a tuft of cottony hairs at one end; fruiting begins in mid-July.

Habitat: infrequent on moist prairies; also on roadsides and in other disturbed places; often growing in patches.

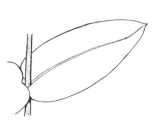

A. sibiricum

Indian hemp
Hemp dogbane
Apocynum cannabinum L.

Apocynum cannabinum is similar to *A. sibiricum* except the leaves have *short leaf stalks*, 1/8″ to 3/8″ long. The leaves are similar in size with a *more distinct point at the tips* and are *rounded* at the bases. The flowers and fruits are similar. Flowering is from late June to early August. Fruiting begins in mid-July. *A. cannabinum* is common in *moist, open woods*, on roadsides, and in open, disturbed places and is also found on mesic prairies. It is less common in northeast Iowa. *A. cannabinum* is sometimes a troublesome weed in minimum-till crop fields.

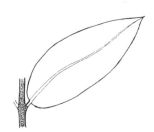

A. sibiricum

A. cannabinum

Spreading dogbane

Apocynum androsaemifolium L.

Stem: perennial; 2 1/2′ to 4′ tall;
  shrublike with spreading branches;
  dark reddish; milky juice in the stem
  and leaves; smooth.

Leaves: opposite; oval with round bases
  and pointed tips; 2 3/4″ by 1 1/2″; leaf
  stalks 1/4″; few hairs above, smooth
  below.

Inflorescence: few, nodding flowers
  on short flower stalks from the stem
  tip and upper leaf axils.

Flowers: pink (sometimes white)
  corolla, with red stripes on the
  inside, bell-like with spreading lobes,
  5/16″ long; calyx 1/16″ long with
  pointed lobes; flowering from mid-
  June to mid-July.

Fruits: seedpods are long, narrow follicles, 3″ to 7″ long by 1/8″ in diameter; seeds with a tuft of cottony hair
  at one end; fruiting begins in
  mid-July.

Habitat: common in open woods;
  also on roadsides and mesic prairies;
  less common in southern and western Iowa.

A. cannabinum   A. sibiricum   A. androsaemifolium

ASCLEPIADACEAE
Milkweed Family

Common milkweed
Asclepias syriaca L.

Stem: perennial, 2′ to 3′ tall; unbranched; hairy; with milky juice in the stem and leaves.

Leaves: opposite; oval with rounded base and tip; 4 1/2″ by 2 1/2″; hairy above and below; leaf stalk 1/4″ to 1/2″.

Inflorescence: many-flowered umbels (up to 100 flowers) from the stem tip and upper leaf axils; on hairy stalks.

Flowers: pink, fragrant flowers, petals 1/4″ long and reflexed, hoods 1/4″ tall with protruding horns; flowering from late June to late July.

Fruits: seedpod is a follicle; 4″ long by 1″ in diameter, tapering to a curved tip; surface warty and covered with fine hairs; seeds 1/4″ by 1/8″, flat, brown with a tuft of white hairs at one end; fruiting begins in late July, and seed release begins about mid-September.

Habitat: frequent in moist to mesic open places with some disturbance; not common on prairies.

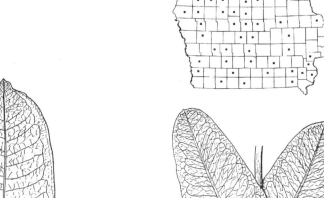

A. syriaca

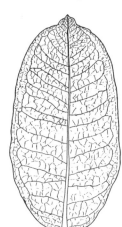

A. syriaca

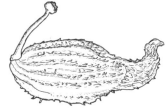

Prairie milkweed
Sullivant's milkweed
Asclepias sullivantii Engelm. ex Gray

Asclepias sullivantii is very similar to *A. syriaca* except it is generally less robust, with *smooth stems, leaves, and fruits*. The leaves are *narrower* and opposite, *6″ by 2″*, and with slightly *chordate bases* and rounded tips with *tiny points*. The upper leaves are held more erect. There are *fewer flowers* per umbel. The fruits are about the same size, but they are *smooth* rather than hairy. Flowering begins in late June and continues until early August. Fruiting begins in late July. Seeds are shed in mid-September. *A. sullivantii* is restricted to *low* to mesic prairies. Prairie milkweed is uncommon on prairies, but perhaps it is often overlooked because of its close resemblance to *A. syriaca*.

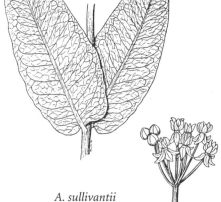

A. sullivantii

Sand milkweed
  Blunt-leaved milkweed
Asclepias amplexicaulis Smith

Asclepias amplexicaulis is very similar
to common milkweed except the stem
is usually *shorter*, about 2′ tall, with
smooth stems and leaves. The leaves are
larger, 5″ by 3″, with *blunt tips* (except
for a small, sharp point at the tip of the
mid-rib), and they have *chordate bases*.
Usually there is *one terminal umbel
with* about *20 flowers*, but there may be
between 10 and 30 flowers. The flowers
have *greenish purple petals* and pink
hoods. Flowering begins in early July,
and seeds are shed in September.
A. amplexicaulis is uncommon on
sandy prairies and is also found on
dry prairies.

Swamp milkweed
Asclepias incarnata L.

Asclepias incarnata is *taller* than common milkweed, up to 5′, with only a
line of hairs on the stem. The *leaves are
long and narrow*, 5″ by 1/2″, *tapering*
at both ends, and are *sparsely hairy*.
Umbels are terminal and from the
upper axils, forming a *corymbiform
inflorescence* of about 30 flowers per
cluster. *Flowers are pink to red. Fruits
are shorter and narrower*, 3″ by 3/8″,
and beaked for one-third the length.
Flowering begins in early July and
continues until early August; fruiting
begins in late July, and seeds are shed
in September. *A. incarnata* is common
on marshes and on lowland prairies.

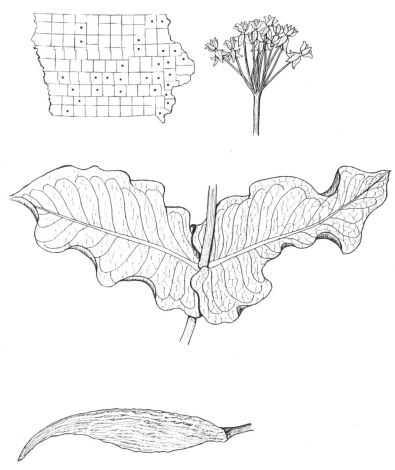

A. amplexicaulis

A. incarnata

Whorled milkweed
Asclepias verticillata L.

Stem: perennial; 2′ to 2 1/2′ tall;
    unbranched; smooth; milky juice in
    the stem and leaves.
Leaves: alternate; crowded; linear, 2″
    by 1/16″; lightly hairy above and
    below.
Inflorescence: short-stalked umbels
    from the upper leaf axils; eight to
    ten flowers per umbel, two to ten
    umbels per plant; umbels about
    1″ across.
Flowers: corolla white, the reflexed
    petals 1/8″ long, hoods and horns
    above the petals 1/8″ long; flowering
    from early July to late August.
Fruits: seedpod is a follicle, 3 1/2″ long
    by 1/4″ in diameter, tapering to both
    ends; dark brown when mature;
    seeds are flat, 3/16″ by 1/4″, with a
    tuft of cottony hairs at one end;
    fruiting begins in early August; seeds
    begin flying in mid-September.
Habitat: common on mesic to dry
    prairies to moist prairies; also com-
    mon on roadsides and in open
    places; often growing in patches.

Butterfly weed
Pleurisy-root
Asclepias tuberosa L.

Stem: perennial; 2 1/2′ to 3 1/2′ tall; unbranched; densely hairy; without milky juice.

Leaves: alternate; elongate with chordate base and tapering tip; 3″ by 1/2″; smooth-edged; hairy above and below.

Inflorescence: umbels from stem-tip and upper (and sometimes middle) leaf axils, producing corymbiform inflorescence; twelve to fifteen flowers per umbel.

Flowers: bright orange with petals 1/4″ long and reflexed, hoods 3/16″ tall; flowering extends from mid-June to mid-August.

Fruits: seedpod is a follicle; 4″ long by 3/8″ in diameter; hairy with a long, tapering beak; seeds are flat, brown, 1/4″ by 3/16″, with a tuft of hairs at one end; fruits become evident in mid-July, and seed release begins in early August; plants often fail to set fruits.

Habitat: infrequent, mostly on dry prairies but also found in the Loess Hills and on moist prairies; often associated with some disturbance.

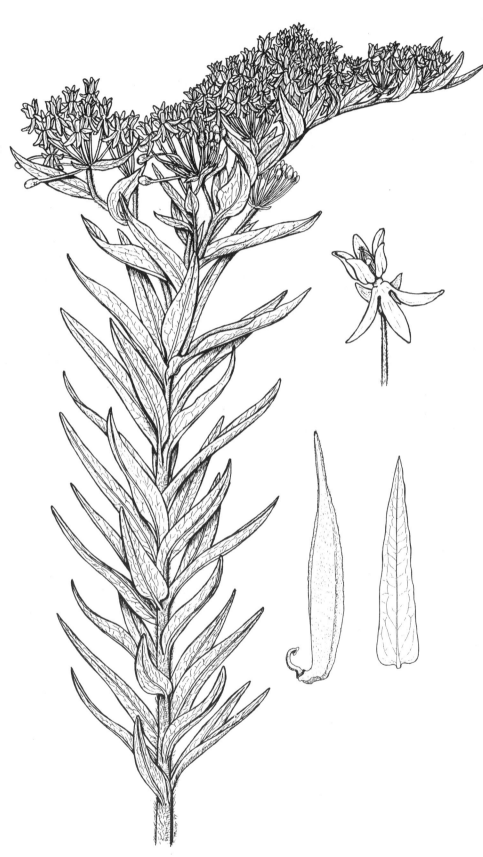

Green milkweed

Green-flowered milkweed
Asclepias viridiflora Raf.

Stem: perennial; 1 1/2′ to 2′ tall;
unbranched; hairy.

Leaves: mostly opposite; oval, tapering
to both ends; 3″ by 1″; sessile; leaf
edges somewhat wavy; sparsely hairy
above, smooth below.

Inflorescence: one to three umbels
from the upper leaf axils; umbels
either sessile or on stalks less than
1/4″ long; up to fifty flowers per
umbel.

Flowers: petals greenish, 1/4″ long and
reflexed, hoods above the petals
3/16″ long, no horn projecting from
the hood; flowering from mid-June
to mid-July.

Fruits: seedpod is a follicle, 3 1/2″ long
by 1/2″ in diameter, hairy; toward
the tip the pod narrows abruptly to
a 1″ beak; seeds are flat, 1/4″ by 3/16″,
with a tuft of cottony hairs at one
end; fruiting begins in late July, and
seeds begin to fly in September.

Habitat: common on mesic and
upland prairies; much less frequent
in eastern Iowa.

Some plants have leaves longer and
narrower than typical for the species.

variations

Tall green milkweed
Asclepias hirtella (Pennell) Woodson

Asclepias hirtella is similar to green milkweed but *taller*, to 3′ tall. The leaves are *long and narrow* (6″ by 1/2″), with *hairs on the veins*. The umbels are *stalked* (1/2″ to 3/4″). There is a narrow *"waist"* in the flower below the hoods, and the fruits *taper to a point*. Flowering occurs from early July to early August; fruiting begins in early August. *A. hirtella* is found *infrequently* on mesic to moist prairies.

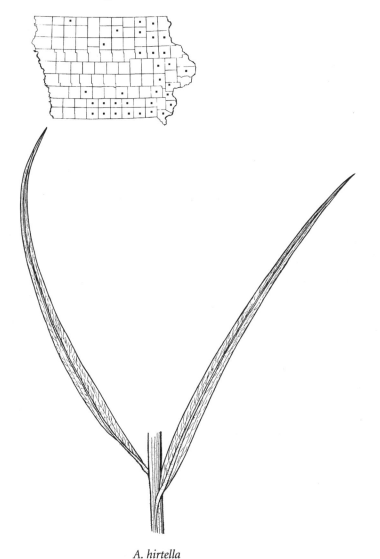

A. hirtella

Oval milkweed
Oval-leaved milkweed
Asclepias ovalifolia Dcne.

Asclepias ovalifolia is similar to green milkweed but with *shorter, wider leaves* (2″ by 1″) which are *finely hairy below*. There are *fewer flowers per umbel* (about fifteen) and the flowers are slightly smaller and *greenish white to greenish purple*. Fruits are *shorter and thicker* (3″ long by 3/4″ in diameter) and *taper gradually* to the tip. Flowering is from mid- to late June, and fruiting begins in mid-July. *A. ovalifolia* is *infrequent* on *dry* prairies.

A. ovalifolia

ASTERACEAE
Daisy Family

Western yarrow
Achillea millefolium L.

Stem: perennial; 2′ to 3′ tall; unbranched; hairy.

Leaves: alternate; highly pinnately divided into narrow segments; 5″ by 1″; hairy.

Inflorescence: corymbiform cluster of many heads.

Heads: 1/8″ in diameter; about five small white rays per head; flowers from early June to mid-July.

Fruits: flat, oval, 1/16″ by 1/64″; fruiting begins in early July.

Habitat: from dry to moist prairie; also in pastures, on roadsides, and in other open places.

A mutant with rose-colored rays is sometimes observed in the field and is available commercially. Apparently, yarrow is only native to the extreme southeastern counties in Iowa.

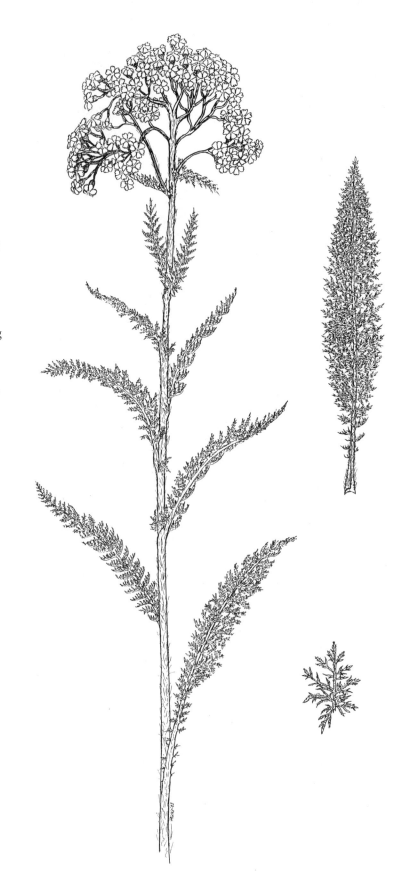

Western ragweed
Ambrosia psilostachya DC.

Stem: perennial; 1 1/2′ tall; branching;
 hairy.

Leaves: alternate; basal leaves deeply
 pinnately lobed; 3″ by 1″; hairy.

Inflorescence: at the tips of branches;
 elongated heads on stalks 2″ long;
 male heads above, a few female
 heads at the base.

Heads: fillaries 1/4″ long in female
 heads obscuring flowers; flowering
 begins in early August and continues until mid-September.

Fruits: top-shaped "seeds" (fruits)
 about 1/8″ long with a central point
 at one end surrounded by several
 other points; fruiting begins in mid-
 September.

Habitat: on dry and sandy prairies.

Pussytoes

Field pussytoes
Antennaria neglecta Greene

Stem: perennial; very short with basal
leaves; flower stalk somewhat lax
and up to 6″ long.
Leaves: basal; widest above the middle
with long-tapering bases and
rounded to sharp points at the tips;
2″ by 1/2″ and smaller; sessile with
one main vein; very hairy above and
below.
Inflorescence: several heads clustered
at the tip of a hairy flower stalk having tiny (1/2″ long) pointed leaves.
Heads: white; 1/2″ across, rays fringed
at the tip; fuzzy when mature with
plumes on the "seeds"; fillaries 1/4″
long with a dark spot near the tip.
Fruits: "seeds" (fruits) 1/32″ long with
1/4″ plumes.
Habitat: common; usually in clusters
of plants; in openings in woods and
on dry and sandy prairies.

Ladies'-tobacco

Plantain-leaved pussytoes
Antennaria plantaginifolia (L.)
Richardson

Antennaria plantaginifolia is similar to
A. neglecta with wider leaves each with
three main veins. The *leaves on flower
stalk are larger* (to 1″ long), the *flower
stalk is longer* (to 1′), and the heads
larger. There is *no spot on the fillaries*.
Flowering is from mid-April to mid-
May, and fruiting begins in early May.
A. plantaginifolia is found commonly,
usually in clusters, and mostly in
openings in woods.

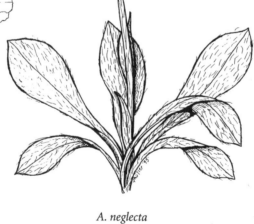

A. neglecta

A. neglecta

A. plantaginifolia

White sage
Artemisia ludoviciana Nutt.

Stem: perennial; 2′ to 3′ tall; sometimes branched toward the tip;
hairy.

Leaves: alternate; elongate with
tapered bases and acute tips; 2″ by
3/8″; few teeth on margin; densely
hairy above and below, lighter green
below; sessile.

Inflorescence: clusters of heads in the
leaf axils near the tips of the
branches.

Heads: green; small, roundish (1/8″
long); no ray flowers; flowering
begins in late August.

Fruits: "seeds" (fruits) tiny, elongate,
about 1/32″ long; fruiting begins in
early September.

Habitat: common on dry to mesic and
sandy prairies and in open places.

This species is a relative of sagebrush
found further west. The crushed leaves
produce the aroma of sage.

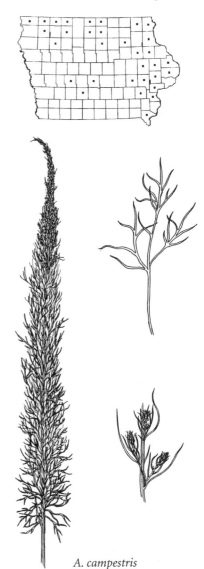

A. ludoviciana

A. ludoviciana

Western sagewort
 Tall wormwood
Artemisia campestris L.
   A. caudata Michx.

Artemisia campestris is a plumelike
plant with *finely divided leaves 2″ to 3″
long*. The upper half of the plant has
heads scattered on short branches. The
flower heads and "seeds" (fruits) are
similar to white sage. Flowering is
from mid-August to mid-September,
and fruiting begins in mid-September.
A. campestris is infrequent on dry and
sandy prairies and in open places and
is often found in disturbed places.

A. campestris

Panicled aster

Aster lanceolatus Willd.
   A. simplex Willd.

Stem: perennial; 2 1/2′ to 4′ tall; more
   or less branched above; sometimes
   hairy; lower leaves falling by flowering time.

Leaves: alternate; linear, tapering bases
   and pointed tips; main stem leaves
   3″ by 3/8″, leaves on branches
   smaller (3/4″ by 1/8″); often serrated;
   smooth above and below; sessile.

Inflorescence: highly branched, leafy,
   elongate.

Heads: white, 3/4″ across; fillaries 1/4″
   tall, green-tipped with a narrow,
   white border; many white rays; yellow disk flowers; flowering from
   early to late September.

Fruits: "seeds" (fruits) 1/16″ long,
   pointed at one end with plumes of
   1/8″ hairs at the other; fruiting
   begins in mid-September.

Habitat: infrequent on low, moist
   prairies and marshes to upland
   prairies.

Heath aster
Frost weed
Aster ericoides L.

Aster ericoides is similar to *A. lanceolatus* except it is *shorter (1′ to 3′)* and more *highly branched above with few hairs* on the stem. The *leaves are smaller (1″ by 1/8″ to 1/2″ by 1/16″)*, with flower heads also *smaller (1/2″ across)* and *crowded on one-sided branches*. *A. ericoides* is highly variable in size and vigor. Flowering is from late August to late September, and fruiting begins in mid-September. *A. ericoides* is common on dry to mesic prairies and is also found on roadsides and in open, disturbed places.

Hairy aster (*A. pilosus* Willd.) has similar heads, but the plant is more robust and weedy and has spreading hairs on the upper stems. Also, the fillaries are inrolled at the tip.

A. ericoides

A. ericoides

Flat-topped aster
White aster
Aster umbellatus Miller

Aster umbellatus usually has unbranched stems and *retains its lower leaves* through flowering. The leaves have *short hairs above and below*. The heads are arranged in a *corymbiform shape* at the top of the stem, and there are *fewer (6–7) ray flowers*. Flowering is from mid- to late August, and fruiting begins in late August. *A. umbellatus* is infrequent on *moist prairies and marshes*.

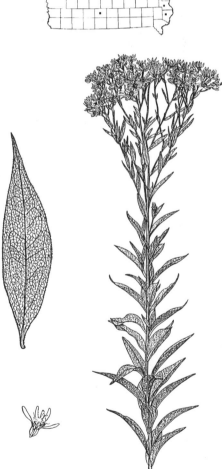

A. umbellatus

Smooth blue aster

Aster laevis L.

Stem: perennial; 2′ to 2 1/2′ tall; unbranched below the inflorescence; smooth.

Leaves: alternate; oval with clasping, heart-shaped bases and long-tapering tips; variable in size, 3″ by 3/4″ and smaller; sometimes toothed; smooth above and below.

Inflorescence: few heads on branched stalks.

Heads: blue rays with yellow disk flowers; about 1″ across; fillaries with a dark green diamond at the tip; flowering from late August to late September.

Fruits: "seeds" (fruits) 1/16″ long with 3/16″ plumes of hairs at one end; fruiting begins in mid-September.

Habitat: frequent on dry to moist prairies; also found in open woods.

Sky-blue aster

Azure aster

Aster azureus Lindley

Aster azureus is *somewhat taller* than
A. laevis, with the lower leaves *on long
leaf stalks* and *somewhat chordate at
their bases* with *a few hairs on the upper
surfaces*. Heads are 1″ across with azure-
blue rays and yellow disk flowers. *A.
azureus* flowers from early to late September, with fruiting beginning in
mid-September. *A. azureus* is frequent
on dry to moist prairies.

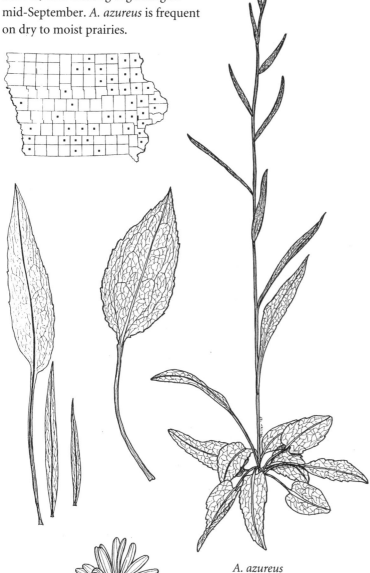

A. azureus

Willow aster

Aster praealtus Poiret

Aster praealtus is similar to *A. laevis* in
height and branching. The *leaves are
sessile* but not clasping, linear, with
long-tapering tips. They are *narrower*
(2″–3″ by 1/4″–5/16″) and much smaller
above (3/4″ by 3/16″) with *prominent
veins below*. The inflorescence is more
branched. The heads have *bluish purple
rays with yellow disks*; the fillaries are
3/16″ long and long-pointed with the
dark mid-rib widening toward the tip.
Flowering is from early to late September, and fruiting begins in late September. *A. praealtus* is infrequent on moist
prairies and in open woods.

A. praealtus

Silky aster
Aster sericeus Vent.

Aster sericeus is *shorter* (1′ to 2′) than
A. lanceolatus, with *wiry stems* and
smaller, sessile leaves (upper 3/8″ by
3/16″) which are *silky above and below.
The lower leaves drop during the flowering season.* In the heads the *rays are
purple to violet.* Flowering is from early
to late September, with fruiting beginning in mid-September. *A. sericeus* is
infrequent on dry, rocky, and sandy
prairies.

A. sericeus

A. sericeus

Aromatic aster
Aster oblongifolius Nutt.

Aster oblongifolius is similar to *A. laevis*, but *shorter*, 6″ to 2′ tall; and *with
wide-spreading branches on the upper
two-thirds of the stem.* The stem is
short-hairy, and the leaves are *smaller*,
1 1/2″ by 3/8″, especially on the side
branches. The leaves are *elongate with
abrupt, tapered bases and pointed tips*
and with *short, stout hairs on the leaf
margins.* The heads are clustered at the
ends of *short branches with leaves 1/4″
by 1/16″*. The heads and fruits are similar. Flowering is from late August to late
September, with fruiting beginning
in mid-September. *A. oblongifolius* is
common on *dry, upland prairies and
Loess Hill prairies* and is less common
in northern and eastern Iowa.

A. oblongifolius

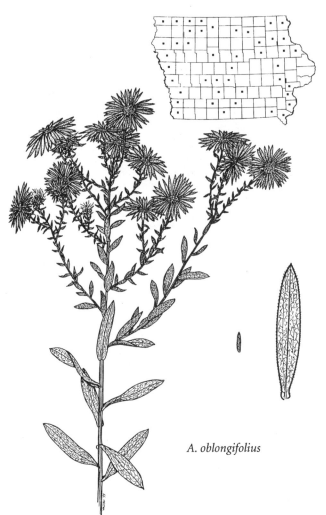

A. oblongifolius

New England aster

Aster novae-angliae L.

Stem: perennial; 2 1/2′ to 4′ tall;
unbranched below inflorescence;
hairy.

Leaves: alternate; sessile with chordate
bases; 2″ by 1/2″ on the lower stem
and gradually decreasing in size
above; hairy below.

Inflorescence: little to much branched.

Heads: rays purple with yellow disk
flowers; head 1″ across; fillaries
sharp-pointed, 1/4″ long; flowering
from late August to late September.

Fruits: "seeds" (fruits) 1/16″ long;
plumes 1/8″ long; in fruit the head
becomes a hairy ball, 1/2″ in diameter; fruiting begins in late September.

Habitat: frequent on moist prairies,
also on roadsides and in open
places.

False boneset

Brickellia eupatorioides (L.) Shinners
  Kuhnia eupatorioides L.

Stem: perennial; 2 1/2′ to 3 1/2′ tall; unbranched; hairy.

Leaves: alternate; crowded on the stems; narrow, with short-tapered bases and long, sharp tips; sessile; 2 1/2″ by 1/2″ and smaller; few to many marginal teeth (three to four per inch); hairy above and below; dark pits on the leaf surface.

Inflorescence: numerous heads on branched flower stalks from the stem tip and upper leaf axils.

Heads: white disk flowers (no ray flowers); heads 3/8″ wide; fillaries hairy, lance-shaped, about 1/4″ tall; flowering from late July to late August.

Fruits: "seeds" (fruits) 1/8″ long with fuzzy plumes; in fruit the head is a fuzzy ball 3/4″ in diameter; fruiting begins in mid-August.

Habitat: on dry and sandy prairies to moist prairies in western Iowa; also on less disturbed roadsides and in open places.

Prairie Indian plantain
Cacalia plantaginea (Raf.) Shinners
  C. tuberosa Nutt.

Stem: perennial; 3′ to 5′ tall; branching near the top; smooth with riblike lines.

Leaves: alternate; mostly on the lower stem and reduced in size above; oval and tapering to both ends, blades 5″ by 3″ with 3″ to 7″ leaf stalks; smooth-edged; smooth above and below.

Inflorescence: corymbiform, with flower stalks from the stem tip and upper leaf axils.

Heads: white; fillaries with prominent ribs; five disk flowers per head; flowering from late June to mid-July.

Fruits: "seeds" (fruits) 3/16″ long with prominent veins and rough surfaces, hairy plumes 5/16″ long; fruiting begins in late July.

Habitat: infrequent on wet to moist prairies, sometimes on drier sites.

Field thistle
Cirsium discolor (Muhl. ex Willd.) Sprengel

Stem: biennial; 4′ to 5′ tall; unbranched; hairy.

Leaves: alternate; deeply, pinnately lobed with sharp points; 12″ long by 2″ wide; sessile; green with few long hairs above and white with dense hairs below.

Inflorescence: few heads from stem tip and upper axils.

Heads: purple disk flowers protrude above 1″ fillaries which are weakly spine-tipped; heads about 1″ in diameter; flowering from mid-August to mid-September.

Fruits: "seeds" (fruits) 3/32″ long; plume hairs finely branched, 3/4″ long; fruiting begins in late August.

Habitat: frequent on upland to lowland prairies, on roadsides, in pastures, and in open places; especially when disturbed.

C. discolor

Bull thistle (*Cirsium vulgare* [Savi] Tenore), a weedy alien, is similar but has leaf edges that continue down the stem as wings and lacks the whitish underside of the leaf.

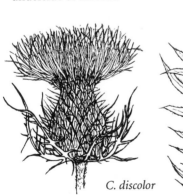

C. discolor

Tall thistle
Cirsium altissimum (L.) Sprengel

Cirsium altissimum is often *taller* than *C. discolor* with *stiff, short hairs on the stems*. The leaves are *less lobed and less spiny*. Flowering is from early August to mid-September, with fruiting beginning in mid-August. *C. altissimum* is *infrequent* on moist to dry prairies and in open, disturbed sites.

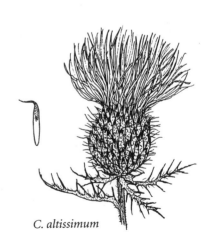

C. altissimum

Flodman's thistle
Cirsium flodmanii (Rydb.) Arthur

Cirsium flodmanii is *perennial* and *shorter* than *C. discolor* (1′ to 2 1/2′ tall) with stems densely white-hairy. The *leaves are smaller* with many small lobes on the margins. The lower surfaces are very densely hairy and somewhat hairy on the upper surface as well. *Heads are a little smaller.* Flowering is from *late June to early August*, and fruiting begins in late July. *C. flodmanii* is common on dry, rocky, and sandy prairies and on Loess Hills prairies.

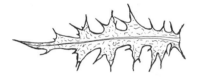

C. flodmanii

Hill's thistle
Cirsium hillii (Canby) Fern.

Cirsium hillii is *perennial and shorter* than *C. discolor* (about 2′ tall). The leaves are *shallowly, pinnately lobed* (8″ by 1 1/2″) with *undulate margins.* The leaves are also *very spiny* and *green on the undersides*. There is usually one or a few *large heads* (2″ across). Flowering is from *mid-June to mid-July*, and fruiting begins in mid-July. *C. hillii* is *infrequent* on upland to sandy prairies.

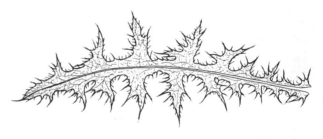

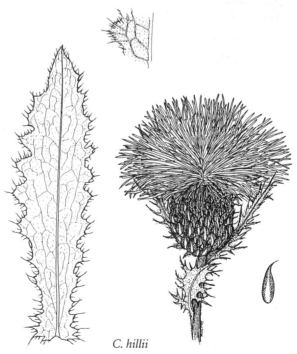

C. hillii

Tickseed

Stiff tickseed

Coreopsis palmata Nutt.

Stem: perennial; 2′ to 2 1/2′ tall; unbranched; smooth.

Leaves: opposite; divided into three elongate lobes; 1 3/4″ by 3/4″ overall; smooth above and below.

Inflorescence: few heads at the ends of leafy flower stalks from the stem tip and upper leaf axils.

Heads: yellow rays with a notch at the tip, yellow disk flowers, the head 1 1/2″ across; fillaries dark yellowish, 3/8″ long; flowering from late June to mid-July.

Fruits: "seeds" (fruits) 3/16″ long, flat and curved; without a plume; only the marginal flowers in the head producing fruits; fruiting begins in early July.

Habitat: frequent on mesic and dry prairies.

Plants are usually found in patches rather than singly.

Pale coneflower

Echinacea pallida Nutt.

Stem: perennial; basal leaves with a 2′
  to 3′ flower stalk; few, long hairs.

Leaves: mostly basal; elongate-oval,
  blades 7″ by 3/4″ with leaf stalks
  from 6″ for basal leaves to 3/4″ for
  stem leaves; parallel veins in the
  blades; bulb-based hairs above and
  below.

Inflorescence: single head at the top
  of a stalk having stiff hairs and few,
  small leaves.

Heads: purple, drooping rays, 1 1/2″
  long; dark purple disk flowers on a
  conical base, the disk about 1″ tall
  and 1″ in diameter; flowering from
  mid-June to mid-July; rays often
  persist until August.

Fruits: "seeds" (fruits) about 1/8″ long,
  squarish and pointed at one end; no
  plume; fruiting begins in late June;
  often fruits persist in the head
  through the winter.

Habitat: infrequent on dry to mesic
  prairies; sometimes on little-
  disturbed roadsides and in open
  places.

Echinacea angustifolia

Some authorities recognize another species, *Echinacea angustifolia* DC., which is very similar to *E. pallida* except with longer, narrower leaves and shorter, spreading rays in the heads. Most Iowa specimens come from the Loess Hills.

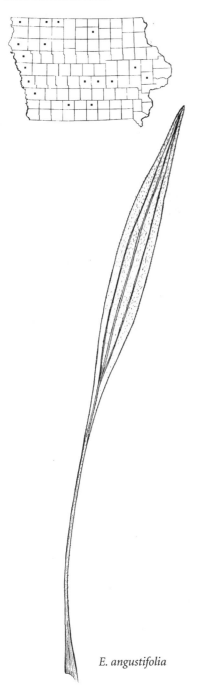

E. angustifolia

Purple coneflower
Echinacea purpurea (L.) Moench

Echinacea purpurea is similar to *E. pallida except the flower stalk is longer* (with small leaves) and is *smooth. The basal leaves are wider* (2 1/2″), and the *margins are toothed.* The ray flowers are shorter and stiffer. Flowering is from early July to mid-August, and fruiting begins in mid-July. *E. purpurea* is very infrequent in woodland edges and prairie openings in southeastern Iowa. It is commonly included in wildflower seed mixtures.

E. purpurea

Fleabane
 Daisy fleabane
Erigeron strigosus Muhl. ex Willd.

Stem: annual; 2′ to 3 1/2′; not branching; slightly hairy.
Leaves: alternate; strap-shaped, tapering to sessile bases, round-pointed
 tips; 1 1/2″ by 1/4″; hairy above and
 below; often slightly toothed.
Inflorescence: branching, from stem
 tip and upper leaf axils.
Heads: many white rays, narrow; head
 3/8″ across; disk flowers yellow;
 resembles the white asters except
 twice as many rays and earlier
 flowering; flowering from early June
 to mid-July.
Fruits: "seeds" (fruits) 1/16″ long with
 plumes 1/8″ long; fruiting begins in
 mid-June.
Habitat: on upland to dry and sandy
 prairies; especially in disturbed sites.

Erigeron annuus (L.) Pers., annual
fleabane, closely resembles *E. strigosus*
except it has larger leaves (3″ by 1″)
with prominent teeth and more ray
flowers. It is more weedy and is often a
pioneer in recently cultivated ground
or other disturbed sites.

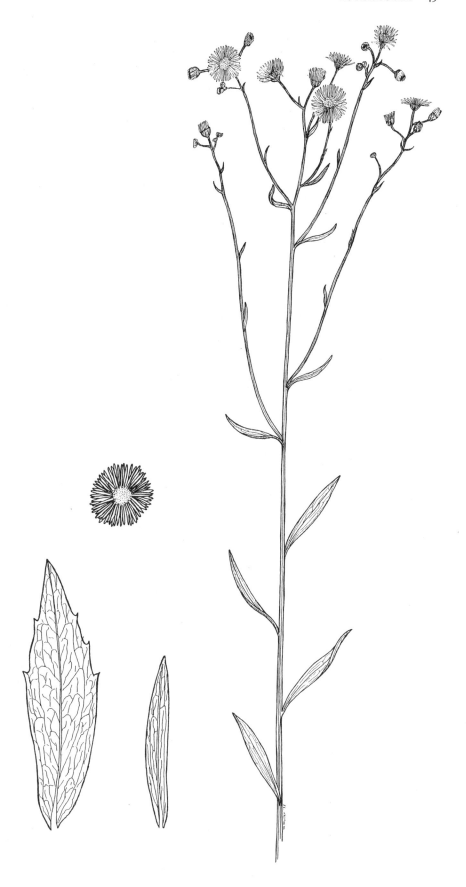

Sneezeweed

Helenium autumnale L.

Stem: perennial; 2 1/2′ to 4′ tall; leaf
 margins continue down stem as
 wings; short-hairy.

Leaves: alternate; crowded; oval with
 tapered bases, sharp tips; 4″ by 1″;
 shallow teeth on margins; hairy
 above and below.

Inflorescence: few heads on branching
 flower stalks from stem tip and
 upper leaf axils; corymbiform.

Heads: few, yellow rays with notched
 ends, 1/2″ to 3/4″ long; head
 2″ across; light brown disk flowers
 form a round head, 1/2″ in diameter;
 few, lance-shaped fillaries; flowering
 from early August to mid-
 September.

Fruits: "seeds" (fruits) 1/32″ long with
 remains of the corolla at the upper
 end, no plume; fruiting begins in
 late August.

Habitat: infrequent on marshes and
 low prairies.

Saw-tooth sunflower

Bigtooth sunflower
Helianthus grosseserratus Martens

Stem: perennial; 4′ to 6′ tall; unbranched; smooth.

Leaves: opposite below, sometimes alternate above; oval with rounded bases and sharp tips; 5″ by 3/4″; tapering to short leaf stalks; large teeth on the margin, six to ten per inch; appressed hairs below, bulb-based hairs above, slightly rough.

Inflorescence: several heads on branching stalks from stem tip and upper leaf axils.

Heads: yellow rays 1″ by 3/8″; head 2 1/2″ across; disk flowers yellow; fillaries lance-shaped with hairy margins, 1/2″ long, reflexed; flowering from mid-August to mid-September.

Fruits: "seeds" (fruits) 1/8″ long; fruiting begins in early September.

Habitat: frequent on moist prairies and in other open sites such as roadsides, less common in drier sites.

H. grosseserratus

Prairie sunflower

Showy sunflower
Helianthus rigidus (Cass.) Desf.
  H. laetiflorus Pers.

Helianthus rigidus is similar to *H. grosseserratus* except the *stem is often rough* and usually *not quite as tall*. Leaves are *smaller, more tapering at the base to a short leaf stalk*, with *smooth or small-toothed margins* and *rough hairs on both upper and lower surfaces*. The *disk flowers are purple*, and the fillaries have *stiff marginal hairs* and are more *blunt-tipped and closely appressed to each other*. Flowering is from mid-August to early September, and fruiting begins in late August. *H. rigidus* is infrequent on dry and sandy prairies and even less frequent in more moist habitats.

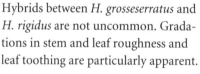

Hybrids between *H. grosseserratus* and *H. rigidus* are not uncommon. Gradations in stem and leaf roughness and leaf toothing are particularly apparent.

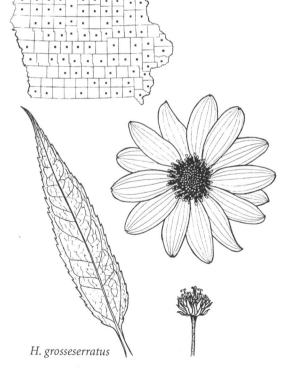

H. grosseserratus

H. rigidus

Jerusalem artichoke
Helianthus tuberosus L.

Helianthus tuberosus is similar to
H. grosseserratus except the *stem is
rough* and the leaves are *shorter and
wider* (4″ by 1 1/2″). The marginal leaf
teeth are not quite as large, and the
leaves are often alternate rather than
opposite in the inflorescence. Usually
there are *more flower heads* forming a
corymbiform inflorescence. The stalks
below the heads are *very hairy*. The fillaries are similar except *H. tuberosus*
has *stiff marginal hairs and appressed
hairs on the lower (outer) surfaces.*
H. tuberosus has *shallow, horizontal
rhizomes with swellings (tubers) several
inches long* and often grows in large
patches. Flowering is from mid-August
to mid-September, and fruiting begins
in late August. *H. tuberosus* is frequently
found on low, moist prairies, or on
roadsides, and in unmanaged places.

bract

Western sunflower

Helianthus occidentalis Riddell

Helianthus occidentalis is generally similar to *H. grosseserratus* but is *shorter*, usually 2′ to 4′ with a *few pairs of leaves, mostly near the base of the stem*. The leaves have *longer leaf stalks, wider blades* (6″ by 3″), *tapered bases, and pointed tips*. The leaves are rough-hairy above and hairy below. There are *fewer heads* on *long, nearly leafless flower stalks*. The *rays are smaller* (3/4″ by 1/8″), the *heads are 1 3/4″ across*, and the *fillaries are oval with sharp tips and stiff marginal hairs*. Flowering is from early August to early September, and fruiting begins in late August. *H. occidentalis* is infrequent on dry and sandy prairies.

The common name obviously does not apply to the distribution in Iowa since it is restricted to the eastern third of the state.

H. occidentalis

H. occidentalis

Maximillian's sunflower

Helianthus maximiliani Schrader

Helianthus maximiliani is generally similar to saw-tooth sunflower *except it is not quite as tall* (4′ to 5′). The leaves are *long and narrow* (5″ to 6″ by 1/2″), *folded along the midrib*, and *rough above and below*. The *flower stalks are hairy*, the *rays are shorter* (3/4″ by 1/4″), and the *heads are 2″* across. The fillaries are lance-shaped and *hairy on both the backs and margins*. Flowering is from late August to mid-September, and fruiting begins in early September. *H. maximiliani* is frequent on dry, upland prairies.

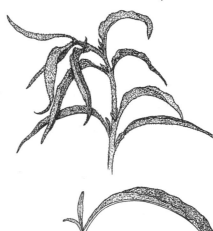

H. maximiliani

Ox-eye
Heliopsis helianthoides (L.) Sweet

Stem: perennial; 2 1/2′ to 4′ tall; branching above; smooth.

Leaves: opposite; ovate with squarish bases and sharp tips; blades 4″ by 2″, on 1/2″ leaf stalks, toothed margins (5–6 per inch), short-hairy above and below.

Inflorescence: heads at the ends of long flower stalks from the stem tip and upper leaf axils.

Heads: yellow rays 3/4″ by 1/4″, heads 2″ across; disk flowers yellow; fillaries widest at the middle, overlapping, short-hairy and with marginal hairs, 5/16″ long; flowering from mid-June to late July; rays retained well into August.

Fruits: "seeds" (fruits) squarish, 1/8″ long, with the remains of the corolla at the upper end, no plume; only the outer flowers of the heads produce fruits; fruiting begins in late June.

Habitat: common on upland to lowland prairies, also sometimes in open woods in eastern Iowa.

Golden aster
Heterotheca villosa (Pursh) Shinners
Chrysopsis villosa Nutt. ex DC.

Stem: perennial; 3′ to 4′ tall; hairy; often several stems from a single root crown.

Leaves: alternate, oval, with tapering bases and pointed tips; 2″ by 3/8″, plants in northwest Iowa have smaller leaves, 3/4″ by 3/32″; hairy above and below.

Inflorescence: single heads at the tips of leafy stems.

Heads: yellow rays and disk flowers; head about 1″ across; flowering over a long period from late June to late August.

Fruits: "seeds" (fruits) 3/16″ long with 1/4″ plumes; fruiting begins after mid-July.

Habitat: very infrequent with a very restricted range in Iowa; on rocky and sandy prairies.

Heterotheca villosa is highly variable with several varieties described. See Gleason and Cronquist (1991) for details.

Hawkweed
Canada hawkweed
Hieracium umbellatum L.

Stem: perennial; 3′ tall; unbranched; smooth; with milky juice in the stem and leaves.

Leaves: alternate; crowded on the stem; oval, tapered to both bases and tips; sessile; 2″ by 3/4″; irregularly toothed; lightly to stoutly hairy above and below.

Inflorescence: branching flower stalks from stem tip and upper leaf axils; hairy; corymbiform.

Heads: only ray flowers; yellow rays, 3/8″ long, with toothed tips; heads 1″ to 1 1/2″ across; fillaries lance-shaped, overlapping, smooth; flowering from early August to late September.

Fruits: "seeds" (fruits) 1/16″ long, dark brown; plumes 1/8″ long; fruiting begins in mid-August.

Habitat: infrequent on moist to upland prairies and in woodland edges.

Eilers and Roosa (1994) recognize *Hieracium canadense* Michx., with longer and finer hairs on the leaves, as a distinct species rarely found in eastern Iowa.

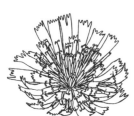

H. umbellatum

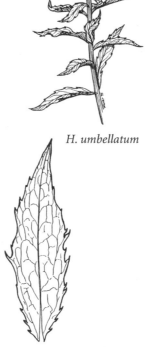

H. umbellatum

Hawkweed
Hieracium longipilum Torrey

Hieracium longipilum is *shorter* than *H. umbellatum* (about 2′). It has *long, fine hairs* (1/2″ long) on the stem and leaves. The leaves are *mostly basal* (4″ by 3/4″) with the stem leaves *1″ long below to 1/4″ long above*. There are few flower heads (3–10) in a corymbiform inflorescence. The heads are similar except the *fillaries are hairy*. The "seeds" (fruits) are *longer* (5/32″) and *black with 1/4″ plumes*. Flowering is from early to late August, and fruiting begins in mid-August. *H. longipilum* is very infrequent on dry, rocky, and sandy prairies.

H. longipilum

False dandelion
Dwarf dandelion
Krigia biflora (Walter) Blake

Stem: perennial; 1 1/2′ tall; smooth; with milky juice in the stem and leaves.

Leaves: basal, 6″ by 1 1/2″, wider above the middle; large, irregular teeth on the margins (similar to dandelion) to smooth margins; one or two clasping leaves on the stem; smooth above and below.

Inflorescence: usually two or three heads on branches at the tip of the stem; sticky hairs on the stem just below the flower heads.

Heads: only ray flowers; yellow rays, 5/8″ long; heads 1″ to 1 1/2″ across; flowering from mid-May to mid-June.

Fruits: "seeds" (fruits) 1/16″ long with fuzzy plumes 1/4″ long; fruiting begins in early June.

Habitat: frequent on moist prairies and in open woods.

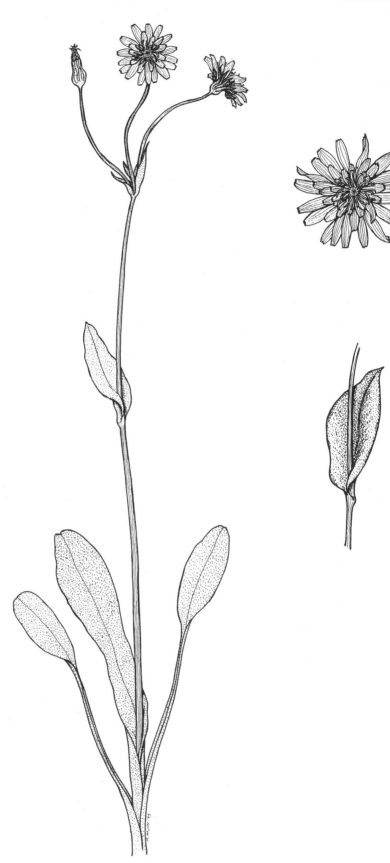

Wild lettuce
Lactuca canadensis L.

Stem: biennial; 3′–6′ tall; unbranched; smooth; with milky juice in the stem and leaves.

Leaves: alternate; deeply, pinnately lobed with five to seven prominent teeth; 6″ by 2″; sessile with small ears at the base; smooth above and below.

Inflorescence: heads on branching flower stalks from the stem tip and upper leaf axils; stalks are short and heads are crowded when flowering begins, but as flowering and fruiting progress the inflorescence becomes the shape of an inverted cone, 1′ to 1 1/2′ long.

Heads: only ray flowers; light yellow rays (sometimes white); head about 3/8″ across; lance-shaped fillaries 5/16″ tall; flowering from early July to mid-August.

Fruits: "seeds" (fruits) flat, circular (1/16″ in diameter) with 1/16″ beaks topped by plumes 3/16″ long; fruiting begins in mid-July.

Habitat: common in disturbed places on prairies, on roadsides, and in open places to open woods; mostly in moist to mesic sites.

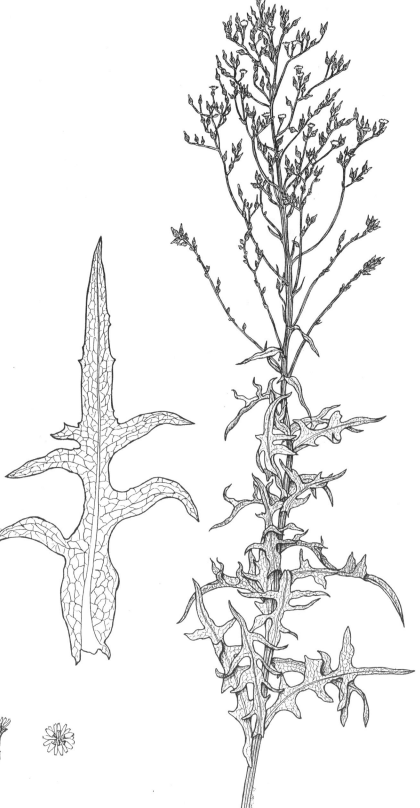

Prairie lettuce
Lactuca ludoviciana (Nutt.) Riddell

Lactuca ludoviciana is similar to *L. canadensis* except the *margins of the leaf teeth are themselves toothed*, sometimes without deep lobing. The ears at the base of the leaves are *larger and pointed*. The inflorescence is the shape of an *upright cone* at maturity. Flower heads are similar in size with the *rays more yellowish*. The *fillaries are longer* (1/2″ to 5/8″). The fruits are similar but *slightly larger*. Flowering is from early July to late August, and fruiting begins in mid-July. *L. ludoviciana* is infrequent on dry prairies and Loess Hills prairies to upland prairies and is usually associated with some disturbance.

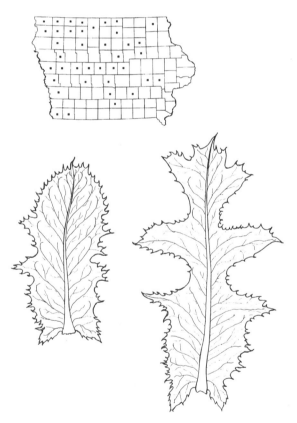

L. ludoviciana

Blue lettuce
Lactuca tatarica (L.) C. A. Meyer
    L. pulchella (Pursh) DC.

Lactuca tatarica is similar to *L. canadensis* except it is *perennial* and is usually shorter (1 1/2′–3′). The leaves have *large, uneven lobes* or are *smooth on the margins* and have *no basal ears*. The inflorescence is similar, but the flowers are *blue*. The heads are *larger* (about 1″ across), the *fillaries are 1/2″ long*, and the "seeds" (fruits) are *larger* (3/16″ in diameter) with a *longer beak* (1/8″). Flowering is from late June to mid-August, with fruiting beginning in mid-July. *L. tatarica* is infrequent on dry prairies and Loess Hills prairies and is sometimes on more mesic prairies.

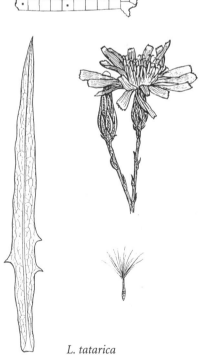

L. tatarica

Prairie blazing star
Liatris pycnostachya Michx.

Stem: perennial; 3′ to 4′ tall; unbranched; scattered stiff hairs on upper half of the stem.

Leaves: alternate; linear, tapering to a pointed tip; lower leaves 12″ by 1/2″, gradually decreasing in size above, to 1/2″ by 1/16″ just below the inflorescence; sessile; smooth above and below; upper leaves with tiny dots on the surface.

Inflorescence: heads all sessile on the upper stem, crowded; usually 4″ to 5″ but up to 10″ long.

Heads: purple disk flowers (no ray flowers) with protruding styles; heads 1/2″ long by 1/4″ across; fillaries about 3/8″ long, the outer (lower) bracts shorter than the inner (upper), with reflexed, abrupt tips with long points and purple margins; flowering from late July to late August.

Fruits: "seeds" (fruits) 1/8″ long, hairy; plumes 5/16″ long; fruiting begins in early August.

Habitat: frequent; mostly on moist prairies but also on drier prairies.

Rough blazing star
Liatris aspera Michx.

Liatris aspera is similar to *L. pycnostachya* in height and general appearance. The leaves are *somewhat shorter*, from 5″ by 1/2″ to 1″ by 1/8″, and long-tapering to leaf stalks on the lower leaves. The heads in the inflorescence are *more widely spaced*, with a leafy bract below each head. The heads are similar except for somewhat *larger fillaries* (5/16″ tall), *closely appressed*, and with *rounded tips*. The fruits are similar. Flowering is from mid-August to mid-September, with fruiting beginning in late August. *L. aspera* is common on upland prairies to dry, rocky, and sandy prairies.

Dotted blazing star
Liatris punctata Hooker

Liatris punctata is *shorter* than *L. pycnostachya* (1′ to 2 1/2′ tall) and often with *several stems from the same root crown*. The *leaves are smaller*, from 4″ by 1/8″ to 2″ by 1/16″, *with tiny dots over both leaf surfaces*. The *inflorescence is shorter*, 3″ to 5″ long, while the *heads are longer*, to 3/4″. The *fillaries are 1/2″ to 5/8″ long, the lower ones with stiff, marginal hairs*. The fruits are similar. Flowering is from early August to early September, and fruiting begins in mid-August. *L. punctata* is frequent on dry, sandy prairies and Loess Hills prairies.

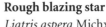

L. pycnostachya

L. pycnostachya

L. aspera

L. punctata

Scaly blazing star
Liatris squarrosa (L.) Michx.

Liatris squarrosa is *shorter* than *L. pyc-nostachya* (1′ to 1 1/2′ tall), with *stiff, spreading hairs* on the stem. The *leaves are shorter* (from 3″ by 1/4″ to 2″ by 3/16″), with *stiff hairs on the lower sur-face* and with the *leaf bases narrowing abruptly*. The *inflorescence is shorter* (2″ to 4″) with *fewer heads*, although they *sometimes overlap*. The heads are larger (5/8″ long), and the fillaries have *long, abrupt, spreading tips (especially the lower) with marginal spreading hairs*. The fruits are similar. Flowering is from late July to late August, and fruiting begins in early August. *L. squarrosa* is infrequent on dry and rocky prairies but does not occur in the Loess Hills.

Cylindric blazing star
Liatris cylindracea Michx.

Liatris cylindracea is *shorter* than *L. pyc-nostachya* (1′ to 2′ tall), with *smooth stems* and *narrower leaves* (to 1/8″). The *heads are stalked* with *only a few heads* (1–3) *to a crowded inflorescence 3″ to 4″ long*. The *heads are longer* (3/4″), with the *fillaries appressed* and *lance-shaped below* to *hair-tipped above*. Fruits are similar. Flowering is from early August to early September, and fruiting begins in mid-August. *L. cylindracea* is very uncommon on dry, rocky, and bluff prairies.

Blazing star
Liatris ligulistylis (A. Nelson) K. Schum.

Liatris ligulistylis is *shorter* than *L. pyc-nostachya* (1′ to 2 1/2′ tall) with a *hairy stem*. The *lower leaves are 7″ by 3/4″* tapering to both ends with a leaf stalk half the length of the blade. The *upper leaves are sessile* and *reduced in size to 1″ by 1/8″*. The *heads are stalked* (1/2″ to 1″). The inflorescence is 4″ to 6″ (12″) long with the heads overlapping. The *heads are wider* (1/2″), and the fillaries are *rounded and overlapping*, 5/8″ long, and *purple-fringed*. The fruits are simi-lar. Flowering is from early August to late August, and fruiting begins in mid-August. *L. ligulistylis* is very uncom-mon on moist to upland prairies.

L. ligulistylis

L. squarrosa

L. cylindracea

Liatris aspera

Liatris punctata

Liatris squarrosa

Liatris cylindracea

Liatris ligulistylis

Rush-pink

Skeleton weed

Lygodesmia juncea (Pursh) D. Don

Stem: perennial; 1′ to 2′ tall; much
branched; smooth; ribbed; with
milky juice in the stems; often with
spherical galls 1/4″ in diameter.

Leaves: alternate; tiny, lance-shaped
(about 1/4″ long); smooth.

Inflorescence: single heads at the tips
of the branches.

Heads: three to five pink ray flowers
per head, 3/16″ long; fillaries in two
rows, the outer (lower) very tiny,
the inner 1/2″ long with darkened
tips; flowering from mid-June to
mid-July.

Fruits: "seeds" (fruits) 1/16″ long;
plumes of many fine hairs, 3/8″
long; fruiting begins in late June.

Habitat: common on Loess Hills prairies and dry, gravelly prairies; often
associated with disturbance.

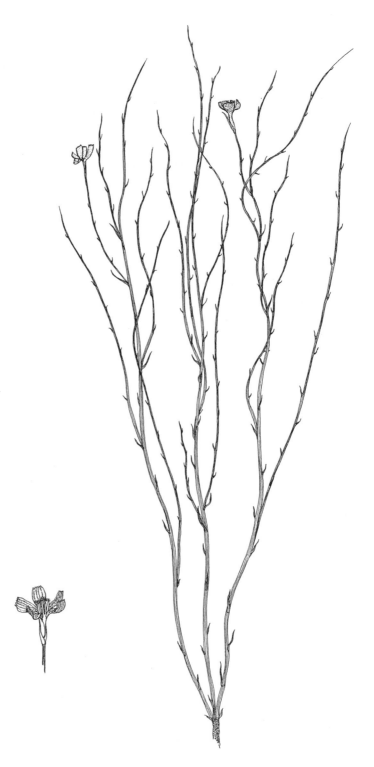

Cut-leaved goldenrod

Machaeranthera spinulosa Greene
    Haplopappus spinulosus (Pursh) DC.

Stem: perennial; 1′ to 2′ tall; often several stems from the single root crown; fine, short hairs.

Leaves: alternate; deeply, pinnately lobed; 1″ by 3/8″ overall; tiny chordate bases; few hairs above and below.

Inflorescence: branching flower stalks from stem tip and upper leaf axils; corymbiform.

Heads: yellow rays, 3/8″ long; head 1″ across; yellow disk flowers; fillaries 1/4″ long with a green diamond at the tip; flowering from late June to mid-August.

Fruits: "seeds" (fruits) 3/32″ long, hairy; plumes 1/4″ long; fruiting begins in late July.

Habitat: frequent on dry, rocky prairies and Loess Hills prairies.

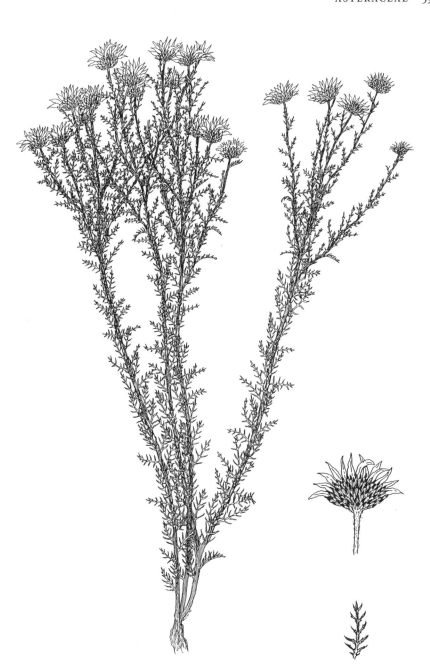

Prairie dandelion

False dandelion

Nothocalais cuspidata (Pursh) Greene

Agoseris cuspidata (Pursh) Raf.

Stem: perennial; very short, basal leaves; with milky juice in the stem and leaves.

Leaves: basal; linear, tapering to a sharp tip, 5″ (to 8″) by 1/2″; smooth with a prominent midrib.

Inflorescence: a single head on a leafless, hairy flower stalk, about 6″ to 12″ tall.

Heads: only ray flowers; ray flowers yellow, heads about 1″ across; fillaries sharp-tipped, 3/4″ long; resembling dandelion except with fewer flowers per head; flowering from early to mid-May.

Fruits: 1/4″ long, spindle-shaped with prominent grooves; plumes of hairs 1/2″ long attached at one end; fruiting begins in mid-May.

Habitat: very uncommon on dry, rocky prairies and Loess Hills prairies.

Feverfew
    Wild quinine
Parthenium integrifolium L.

Stem: perennial; 2′ to 3′ tall;
    unbranched but several stems often
    emerging from a single root crown;
    fine or stiff hairs.

Leaves: alternate, oval with tapered
    bases and pointed tips; lower leaves
    with a long leaf stalk (blade 6″ to 8″
    by 3″), upper leaves smaller and ses-
    sile to clasping; margin of uneven
    teeth about 1/16″ to 1/8″ long,
    approximately eight per inch; with
    few stiff hairs above, smooth below.

Inflorescence: highly branched flower
    stalks from the stem tip and upper
    leaf axils; corymbiform.

Heads: few (five) short, white ray
    flowers with notched tip; head 1/4″
    in diameter; disk flowers white, very
    hairy; fillaries oval, overlapping,
    very hairy, 1/8″ tall; flowering from
    late June to late July.

Fruits: "seeds" (fruits) 3/16″ long;
    plumes 3/8″ long; only the outer
    flowers of the head producing fruits;
    fruiting begins in mid-July.

Habitat: infrequent on upland prai-
    ries; also found on sandy, bluff, and
    moist prairies.

Rough white lettuce
  White lettuce
Prenanthes aspera Michx.

Stem: perennial; 3′ to 4′ tall;
  unbranched; hairy, especially the
  upper third; with milky juice in the
  stem and leaves.

Leaves: alternate; oval with clasping
  bases (sessile) and pointed tips;
  4 1/2″ by 1 1/2″ and smaller; with
  occasional teeth (1/16″ long); with
  short, stiff hairs above and below.

Inflorescence: two or three heads on
  short (1/4″) stalks from the upper
  leaf axils; total inflorescence
  4″ to 8″ long; cylindric.

Heads: only ray flowers; ray flowers
  cream-colored, 5/8″ across; outer
  fillaries short, inner 3/8″ tall, long-
  hairy; flowering from mid-August
  to early September.

Fruits: "seeds" (fruits) 1/4″ long,
  ribbed; plumes 5/16″ long; fruiting
  begins in late August.

Habitat: very uncommon on dry to
  moist prairies.

Glaucous white lettuce

Smooth white lettuce

Prenanthes racemosa Michx.

Stem: perennial, 3′ to 4′ tall;
unbranched; smooth; with milky
juice in the stem and leaves.

Leaves: alternate; wider above the
middle, tapering bases clasping the
stem (sessile), rounded tips, the
upper leaves having small ears at the
base; 10″ by 3″ below to 3″ by 1 1/2″
above; small, irregular teeth on the
margin; smooth above and below.

Inflorescence: ten to fifteen heads on
short, branching stalks from the
upper leaf axils; the total inflorescence 8″ to 12″ long; cylindric.

Heads: only ray flowers; ray flowers
pink to purple; head 5/8″ across;
fillaries overlapping, lance-shaped,
1/2″ tall, long-hairy; flowering from
late August to late September.

Fruits: "seeds" (fruits) 1/4″ long;
plumes 5/16″ long; fruiting begins in
early September.

Habitat: infrequent on moist to
upland prairies, also in open woods.

Gray-headed coneflower
Yellow coneflower
Ratibida pinnata (Vent.) Barnh.

Stem: perennial; 3′ to 5′ tall; branched above; short, appressed hairs.

Leaves: alternate; deeply, pinnately lobed; blades of lower leaves 7″ by 4″ with 4″ leaf stalks, upper leaves smaller and nearly sessile; mostly smooth margins; hairy above and below.

Inflorescence: few heads at the ends of long flower stalks.

Heads: several drooping, yellow rays, 1″ long; brown rounded disk 1/2″ tall and 3/8″ wide; fillaries lance-shaped, 1/4″ long, hairy; flowering from early July to mid-August.

Fruits: "seeds" (fruits) 1/16″ long, black, no plume; fruiting begins in mid-July.

Habitat: common on moist to dry prairies and in woodland edges, also on roadsides and open places; often associated with some disturbance.

R. pinnata

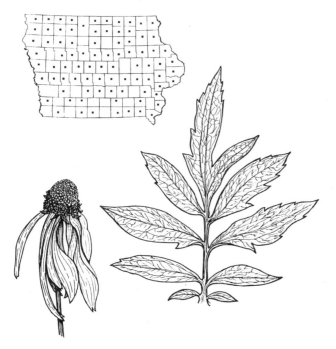

R. pinnata

Long-headed coneflower
Thimble-weed
Ratibida columnifera (Nutt.) Wooton and Standley

Ratibida columnifera is similar to gray-headed coneflower except *shorter* (1′ to 2′) with the stem *branching near the ground*. The *leaves are smaller* (2 1/2″ by 1 1/2″) with *narrower segments*. The ray flowers are *shorter* (3/4″), sometimes with a tinge of red. The disk is *elongate* (1″ by 1/4″) and increases in length to *1 1/2″ in fruit*. Flowering is from early to late July, and fruiting begins in mid-July. *R. columnifera* is common on dry prairies and Loess Hills prairies in western Iowa and very uncommon in eastern Iowa on sandy soil. A commercial variety of *R. columnifera* is known as Mexican hat.

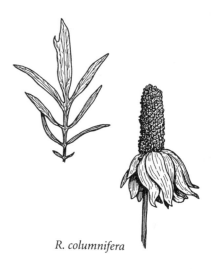

R. columnifera

Black-eyed Susan
Rudbeckia hirta L.

Stem: biennial; 1′ to 2′ tall; branching above the middle; with long-spreading hairs.

Leaves: alternate; oval and widest above the middle, pointed tips, tapering to sessile bases; 3″ by 3/4″ and smaller; spreading hairs above and below.

Inflorescence: several to many heads at ends of branches.

Heads: few, yellow ray flowers, 1″ by 1/4″, held rigidly perpendicular to the axis of the head; disk flowers brown, producing a cone-shaped head (more pronounced in fruit), rays are held until after flowers have produced seeds; flowering from mid-June to mid-July.

Fruits: "seeds" (fruits) 1/16″ long, black, no plume; falling soon after maturing; fruiting begins in late June.

Habitat: common on dry to moist prairies; also on roadsides and in open woods; in greatest abundance in somewhat disturbed places.

Fragrant coneflower
Sweet coneflower
Rudbeckia subtomentosa Pursh

Stem: perennial; 3′ to 5′ tall; unbranched; hairy.

Leaves: alternate; usually deeply three-lobed, the two lateral lobes smaller; 3″ by 2″ with 1/2″ leaf stalks; toothed margin, especially the center lobe; stiff-hairy above and below.

Inflorescence: corymbiform with single heads at the ends of branches from the upper leaf axils.

Heads: few yellow rays, widely spreading; brown disk flowers forming a rounded head; bract around each disk flower hairy near the tip; fillaries overlapping, lance-shaped, hairy, 1/4″ long; flowering from late July to early September.

Fruits: "seeds" (fruits) 3/32″ long, black; fruiting begins in early August.

Habitat: infrequent on moist prairies to woodland edges.

Several other species of *Rudbeckia* resemble *R. subtomentosa*. The best characteristic to distinguish fragrant coneflower is the hairy bract around each disk flower. Other *Rudbeckia* species (except *R. hirta*) are more often found in shady places and have leaves more consistently divided.

Prairie ragwort
Senecio plattensis Nutt.

Stem: biennial; 1′ to 2′ tall; unbranched; tufted, woolly hairs.

Leaves: alternate; basal leaves oval with long leaf stalks, stem leaves sessile, progressively smaller, and more pinnately divided up the stem; basal leaves 1 1/2″ by 3/4″ with 1″ to 4″ leaf stalks; lower stem leaves 3″ long, upper, 1 1/2″ long; toothed margins; hairy, especially the upper leaves.

Inflorescence: tight clusters of heads on branching flower stalks from the upper leaf axils and stem tip; flower stalks woolly.

Heads: numerous (about 12), yellow rays, 1/8″ long; yellow disk flowers; head 1/2″ across; fillaries 1/4″ long; flowering from mid-May to early June.

Fruits: "seeds" (fruits) 1/16″ long with stiff, very short hairs on the angles; plumes of fine hairs, 1/4″ long; fruiting begins in late May.

Habitat: frequent on dry to moist prairies; in open woods, and on roadsides.

Another prairie ragwort, *Senecio pauperculus* Michx., less woolly and longer-lived than *S. plattensis*, is found on moist prairies.

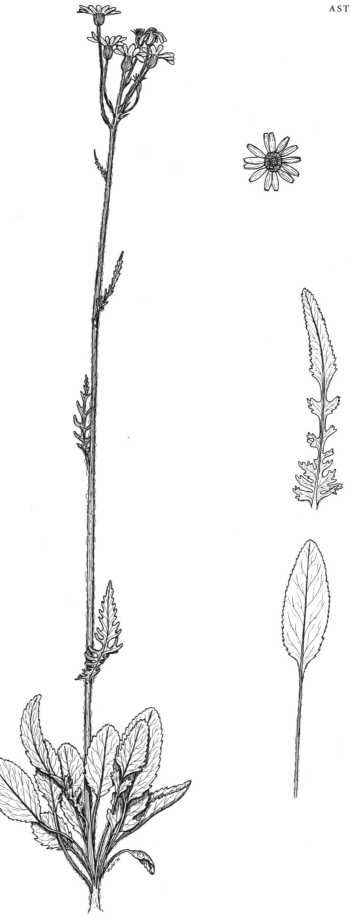

Rosinweed
Silphium integrifolium Michx.

Stem: perennial; 3′ to 5′ tall; unbranched; smooth.

Leaves: opposite; oval with rounded bases and pointed tips; sessile; 4″ by 2″ or narrower to 3 1/2″ by 1″; margins sometimes with small teeth; rough with hairs above and below; often "cupped."

Inflorescence: one to a few flower stalks from the stem tip and upper leaf axils; heads on short, hairy stalks; five to ten heads per plant.

Heads: pale yellow rays 3/4″ by 1/4″; yellowish green disk flowers; head 2″ across; fillaries oval and pointed, 1/2″ tall, hairy; flowering from mid-July to late August.

Fruits: "seeds" (fruits) flat, roundish, 3/8″ long; no plume; only the outer flowers setting fruit; fruiting begins in early August.

Habitat: frequent on dry to moist prairies; also on roadsides and sometimes in open woods.

A variety, *laeve* T. & G. (*Silphium speciosum* Nutt.) (not recognized by Eilers and Roosa 1994), is found in several border counties in southwest Iowa. It is not hairy on the stem or fillaries, the leaves are less hairy, and the leaf margins are smooth.

Compass plant

Silphium laciniatum L.

Stem: perennial; 4′ to 6′ tall; often several stems from one rootstock; hairy early in the season, somewhat woody late in the season.

Leaves: alternate; deeply, pinnately lobed with five to seven pairs of lobes; blades 12″ to 24″ by 6″ with leaf stalks on the lower leaves; margins of the lobes sometimes with large teeth; rough above and below.

Inflorescence: one or two flower stalks with numerous heads widely scattered; heads on short, hairy stalks, 1/2″ to 1″ long.

Heads: yellow rays 1″ by 1/4″; disk 1″ to 1 1/2″ across, the flowers yellow; head 3″ across; fillaries oval with sharp, spreading tips, 1″ long with hairs on margins and outer surfaces; flowering from early July to early August.

Fruits: "seeds" (fruits) flat, oval, concave; about 3/8″ long; no plume; only the outer flowers setting fruit; fruiting begins in mid-July.

Habitat: infrequent on upland prairies to dry, sandy, and wet prairies.

Compass plant is named for its tendency to orient its leaves facing east and west, perhaps as a cooling mechanism on hot summer days. As a result the plant gives a rough idea of the north-south axis but no clue as to which direction is north and which is south.

Missouri goldenrod
Solidago missouriensis Nutt.

Stem: perennial; 1 1/2′ to 2 1/2′ tall; unbranched; smooth.

Leaves: alternate; narrow, tapered bases, sharp tips; 3″ by 3/8″ to 1″ by 1/8″ below the inflorescence; toothed toward the tips; smooth above and below.

Inflorescence: branching flower stalks at the stem tip; many small heads attached to one side of flower stalks; inflorescence often cone-shaped at maturity.

Heads: yellow ray and disk flowers; head 1/4″ across; fillaries 1/8″ tall; flowering from mid-July to mid-August.

Fruits: "seeds" (fruits) 1/32″ long; plumes 1/8″ long; fruiting begins in late July.

Habitat: frequent on dry, upland, and sandy prairies.

Two characteristics that distinguish Missouri goldenrod are its early blooming, the first of the goldenrods, and its lack of hairs on the stem and leaves. Not all stems produce flowers.

flowerless stem,
vegetative stalk

Tall goldenrod
Canada goldenrod
Solidago canadensis L.
Includes *S. altissima* L.

Stem: perennial; 2 1/2′ to 3 1/2′ tall; unbranched; upper part of the stem hairy.

Leaves: alternate; oval with tapering bases and sharp tips; sessile; 3″ by 1/2″; toothed margins; short hairs above and below, somewhat rough; lower leaves drop before flowering.

Inflorescence: branching flower stalk producing cone-shaped or inverted cone-shaped inflorescence; small heads crowded on one side of hairy, curved flower stalks.

Heads: ray and disk flowers yellow; entire head 3/16″ across; fillaries 1/16″ long; flowering from early August to mid-September.

Fruits: "seeds" (fruits) 1/16″ long; fruiting begins in mid-August.

Habitat: common in open, grassy areas, in open woods, and on prairies; quite weedy and increases in untended gardens and fields for several years after abandonment.

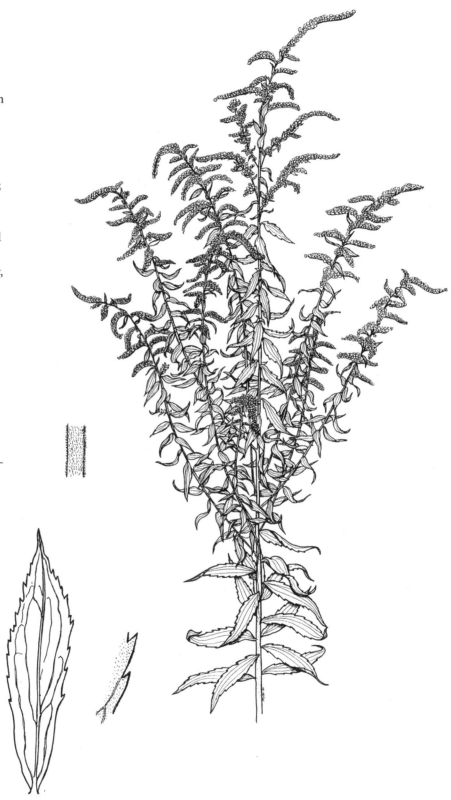

Showy goldenrod
Solidago speciosa Nutt.

Solidago speciosa is similar to tall gold-enrod except the stem is hairy *only in the inflorescence* and the lower leaves are *retained on the stem*. The leaves *gradually diminish in size from bottom to top* (4″ by 1/2″ to 1/2″ by 3/32″) and have *smooth margins*. The inflorescence is *more cylindrical* although it is usually widest at the base. The fillaries are *pointed to rounded*. Flowering is from late August to early October, and fruiting begins in mid-September. *S. speciosa* is infrequent on dry to mesic prairies and sometimes in open woods.

Smooth goldenrod
Late goldenrod
Solidago gigantea Aiton

Solidago gigantea is similar to *S. canadensis* except the stem is *smooth and usually reddish* and is *hairy only in the inflorescence*. There are often several stems from one root crown. The leaves are about the same size and toothed but are *smooth except for fine hairs on the margins*. Flower heads and fruits are similar. Flowering is from early August to mid-September, and fruiting begins in mid-September. *S. gigantea* is common on moist to mesic prairies, in open woods, and in disturbed places.

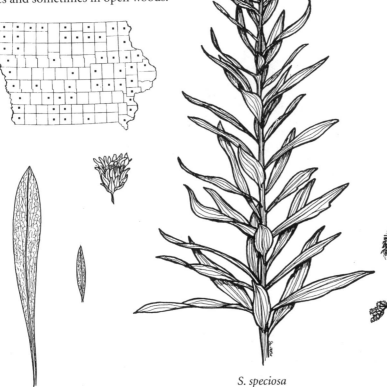

S. speciosa

S. gigantea

Field goldenrod

Gray goldenrod

Solidago nemoralis Aiton

Stem: perennial; unbranched; 1′ to 3′ tall; sometimes more than one stem per root crown; lightly hairy.

Leaves: alternate; basal and lower stem leaves wider above the middle with pointed tips and long-tapering bases; 4″ by 1/2″ below to 2″ by 3/8″ at mid-stem; margins with teeth toward the tip; rough above and below.

Inflorescence: usually cylindrical, sometimes cone-shaped or curved, 4″ to 6″ long; heads crowded on one side of the flower stalks; hairy.

Heads: yellow ray and disk flowers; heads 1/8″ across; fillaries 3/16″ long, overlapping; flowering from mid-August to early September.

Fruits: "seeds" (fruits) 1/16″ long; plumes 1/8″ long; fruiting begins in mid-August.

Habitat: frequent on dry, rocky prairies and in prairie openings in dry woodlands.

Stiff goldenrod
Solidago rigida L.

Stem: perennial; 3′ to 4′ tall;
unbranched except sometimes one
or two branches from the upper
half; finely hairy.

Leaves: alternate; lower leaves with
long leaf stalks, blades 6″ by 2″;
middle and upper leaves sessile,
3 1/2″ by 3/4″ and smaller, tapering
bases and pointed tips; margins
round-toothed; hairy above and
below.

Inflorescence: flower stalks branch to
form a corymbiform inflorescence;
hairy.

Heads: yellow ray and disk flowers;
heads 3/8″ across; fillaries blunt-
tipped, 3/16″ long, hairy; flowering from mid-August to mid-
September.

Fruits: "seeds" (fruits) 1/16″ long,
plumes 3/16″ long; fruiting begins in
late August.

Habitat: common on dry to mesic
prairies to moist prairies; also in
open woods.

Lance-leaved goldenrod

Grass-leaved goldenrod
Euthamia graminifolia (L.)
Nutt. ex Cass.
   Solidago graminifolia (L.) Salisb.

Stem: perennial; 2′ to 3′ tall; smooth.
Leaves: alternate; linear with sharp
   tips; 3″ by 1/4″ to 1″ by 3/16″ above;
   hairy on margins and veins.
Inflorescence: branching flower stalks
   from stem tip and leaf axils in upper
   half to fourth of the stem; producing
   wide, corymbiform inflorescence.
Heads: yellow ray and disk flowers;
   head 1/8″ across; fillaries 1/4″ long;
   flowering from mid-August to mid-
   September.
Fruits: "seeds" (fruits) 1/16″ long;
   plumes 1/8″ long; fruiting begins in
   late August.
Habitat: infrequent on moist (to dry)
   prairies; often found in patches in
   moist roadside ditches.

Riddell's goldenrod

Narrow-leaved goldenrod
Solidago riddellii Frank ex Riddell

Solidago riddellii is similar to *Euthamia
graminifolia* except the leaves are *longer
and wider* (6″ by 1/2″ below to 2″ by
3/8″ above). The inflorescence is similar in shape but *smaller*, and the fillaries are *shorter* (3/16″) and *round-
tipped*. Flowering is from mid-August
to mid-September, and fruiting begins
in late August. *S. riddellii* is found in
wetter habitats, on wet prairies to
marshes.

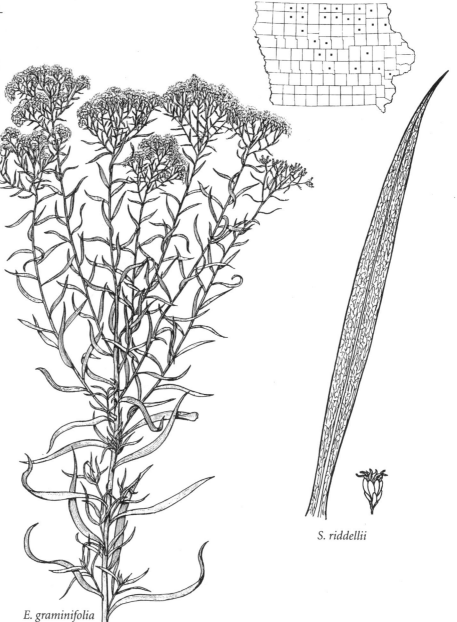

E. graminifolia

E. graminifolia

S. riddellii

Ironweed
  Western ironweed
  Vernonia fasciculata Michx.

Stem: perennial; 3′ to 4′ tall;
  unbranched; smooth.
Leaves: more or less opposite; sessile;
  narrow-oval with tapering bases and
  long-tapering tips; 6″ by 3/4″ below
  to 3″ by 5/16″ above; small teeth
  (10 per inch) on the margins;
  smooth below, sparse-hairy above.
Inflorescence: corymbiform from
  branching flower stalks from the
  stem tip and upper leaf axils; heads
  tightly clustered.
Heads: purple ray and disk flowers;
  entire head 3/8″ across; fillaries
  round-tipped, outer shorter than
  inner, some with spreading marginal hairs, purple; flowering from
  late July to early September.
Fruits: "seeds" (fruits) 1/8″ long with
  prominent ridges; plumes 1/8″ long,
  purple; fruiting begins in early
  August.
Habitat: common on moist to wet
  prairies and in marshes; also in
  open, alluvial woods.

Baldwin's ironweed
Vernonia baldwinii Torrey

Vernonia baldwinii is similar to *V. fasciculata* except the stem is *hairy*, the leaves are *wider and shorter* (4″ by 1″ below to 2 1/2″ by 1/2″ above) and *hairy below*. The inflorescence, heads, and fruits are similar, but the *fillaries are sharp-tipped and reflexed*. Flowering is from mid-July to late August, and fruiting begins in late July. *V. baldwinii* is frequent on *dry* upland prairies and in pastures to open woods.

BORAGINACEAE
Forget-me-not Family

Hoary puccoon
Lithospermum canescens (Michx.)
Lehm.

Stem: perennial; 6″ to 12″ tall; branching above; hairy.

Leaves: alternate; linear, tapered bases
and rounded tips; 2″ by 1/4″;
smooth-edged; sessile; hairy above
and below.

Inflorescence: coiled at the tip,
straightening as flowers develop;
bract below each flower.

Flowers: petals yellow-orange; corolla
tube 3/8″ long with 1/4″ lobes; sharp-
pointed sepals, 3/16″ long, hairy;
flowering extends from early May
to mid-June.

Fruits: very hard, cream-colored nutlets in clusters of four within the
calyx, 1/16″ long; fruiting begins in
late May.

Habitat: frequent on dry to moist
prairies.

Fringed puccoon

Narrow-leaved puccoon
Lithospermum incisum Lehm.

Lithospermum incisum is about the
same height as *L. canescens* and is *less
hairy* on stems and leaves. The *leaves
are smaller*, 1 1/2″ by 1/8″. Flowers are
pale yellow, and the *corolla tube is
much longer* (1″) with *fringed lobes*
1/4″ long. The sepals are sharp-pointed
and *1/2″ long*, and the fruits are similar.
Flowering from early to late May, and
fruiting begins in late May. *L. incisum*
is frequent on *rocky and sandy prairies*.

Hairy puccoon

Lithospermum caroliniense (Walter)
MacM.

Lithospermum caroliniense is *taller*
than *L. canescens*, 2′ to 2 1/2′, and is
more branched. The leaves are 1 1/4″ by
3/8″, tapering to both ends. The *corolla
is golden yellow* and *3/4″ long*. The *calyx
and fruits are also larger*. Flowering is
from mid-May to late June, and fruiting begins in early June. *L. caroliniense*
is infrequent in *very sandy soil*, often
with some disturbance such as roadsides.

False gromwell
Onosmodium molle Michx.

Stem: perennial; to 3′ tall; sometimes
branching above; rough-hairy; often
with several to many stems from
the same root crown, appearing
bushlike.

Leaves: alternate; oval, tapering to both
ends; 2 1/2″ by 3/8″; hairy, especially
on the veins on the upper surfaces;
sessile.

Inflorescence: one-sided, coiled at the
tip and straightening as flowers
develop; at the tips of the stems; with
a leaflike bract below each flower;
hairy.

Flowers: corolla white, tubular with
sharp-pointed lobes; corolla 1/2″ to
3/4″ long; style exceeding the corolla;
flowering from mid-June to mid-
July.

Fruits: nutlets about 1/8″ long, top-
shaped, hard and shiny, four per
flower within the calyx; fruiting
begins in mid-July.

Habitat: common to frequent in
western Iowa to very infrequent
in eastern Iowa; on dry or sandy
upland prairies.

BRASSICACEAE
Mustard Family

Spring cress
Cardamine bulbosa (Schreber) BSP.

Stem: perennial; 1′ to 2′ tall; unbranched; smooth.

Leaves: alternate; lower leaves with leaf stalks, blades 1 1/2″ by 3/4″, without lobes; upper leaves sessile, 3/4″ by 3/8″, smaller above; shallow-lobed; smooth above and below.

Inflorescence: raceme; continues to produce new flowers at the tip.

Flowers: four white petals, with rounded tips, 5/16″ long; four sepals, shorter; flowering from early to late May.

Fruits: seedpod slender, 1 1/4″ by 1/16″; opening into two linear compartments; on a stalk 1 1/4″ long; spaced about 3/8″ apart at maturity; fruiting begins in mid-May.

Habitat: infrequent in moist soil in open woods and on wet prairies.

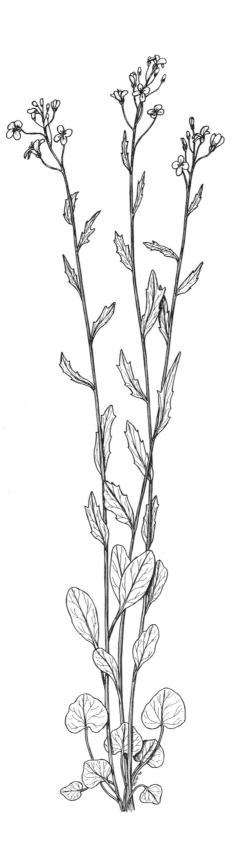

CACTACEAE
Cactus Family

Eastern prickly pear
Opuntia humifusa (Raf.) Raf.
　O. compressa (Salisb.) J. P. Macbr.

Stem: joints flattened into fleshy, oval
　segments about 4 1/2″ long, dull
　green; with 1/2″ (1″) spines solitary
　or in pairs.
Inflorescence: few flowers attached to
　the upper, curving edge of terminal
　joints.
Flowers: many yellow petals, about
　1″ long, many stamens and a single
　style; ovary tapering, petals attached
　at the top of the ovary (inferior
　ovary), with spines; flowering from
　mid-June to mid-July.
Fruits: red, columnar, 1″ to 2″ long.
Habitat: very infrequent; only on dry,
　rocky, or sandy prairies.

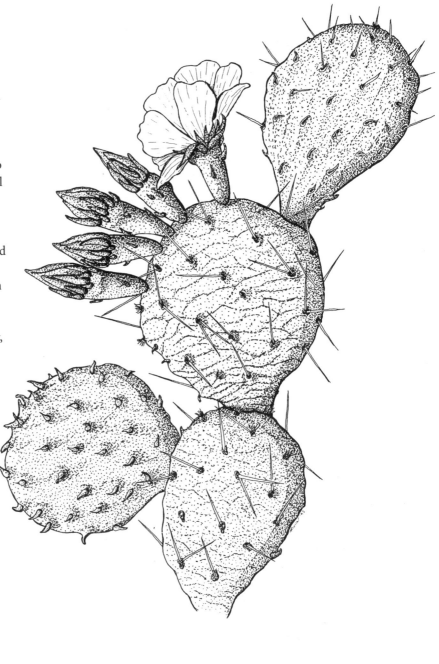

Little prickly pear
Opuntia fragilis (Nutt.) Haw.

Opuntia fragilis is similar to *O. humifusa* except the *joints are smaller*, 1″ long by 3/4″ wide, and *more cylindrical*, and the *spines are shorter*, 1/2″. The flowers are *slightly smaller* with 3/4″ petals, and the fruits are also smaller. Flowering is from mid- to late June, and fruiting begins in early July. In Iowa, *O. fragilis* is found only on Sioux Quartzite (Pipestone) outcrops in extreme northwest Iowa.

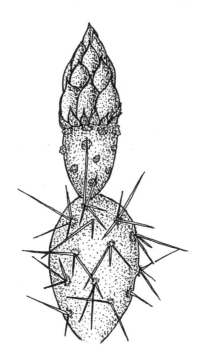

CAMPANULACEAE
Harebell Family

Marsh bellflower
  Bedstraw bellflower
  Campanula aparinoides Pursh

Stem: perennial; sprawling, from 6″ to 2′ long; often with tiny, hooked hairs on the angles.

Leaves: alternate; linear with tapering bases and pointed tips; tiny barbs on the margins; sessile; basal leaves 1 1/2″ by 1/4″, gradually decreasing in size toward the tip of the stem.

Inflorescence: single flowers from upper leaf axils, on stems 2″ to 4″ long with several tiny leaves.

Flowers: light blue; spreading corollas, 1/4″ to 1/2″ long; flowering from early July to mid-August.

Fruits: capsules opening at the tips by tiny points; 1/8″ long; fruiting begins in early August.

Habitat: moist lowland prairies to marshes.

Spiked lobelia

Pale-spiked lobelia

Lobelia spicata Lam.

Stem: perennial; 2′ to 2 1/2′ tall; unbranched; few hairs.

Leaves: alternate; wider above the middle with pointed or rounded tips and tapering bases, 2 1/2″ by 1″; margins slightly toothed; gradually decreasing in size toward the tips; hairy.

Inflorescence: long raceme with leaflike bracts below each flower.

Flowers: light to medium blue, 1/4″ to 1/2″ tubelike corolla with three lower lobes; flowering from early June to mid-July.

Fruits: capsules opening at the tips; 1/4″ long; maturing within the calyx lobes; fruiting begins in late June.

Habitat: frequent on dry to moist prairies, on roadsides, and in little-disturbed open areas.

Great lobelia
   Giant lobelia
Lobelia siphilitica L.

Lobelia siphilitica is *more robust than
L. spicata*, 1 1/2′ to 2′ tall, with spreading hairs on the stem angles. The leaves
are 2″ by 1/2″ to *5″ by 1″* (in wet areas),
widest at the middle, and tapering to
both ends. The leaf margins have *more
prominent teeth* and *sharper tips*, and
there are short hairs on the leaf margins. The *flowers are dark blue, 1″ long*,
with three prominent lower lobes. The
fruits are widest below the middle and
taper to sharp points. Flowering is from
early August to mid-September, and
fruiting *begins in mid-August. L. siphilitica* is frequent on *wet prairies, on
stream banks*, and *in sandy marshes*.

CARYOPHYLLACEAE
Pink Family

Sleepy catchfly
Silene antirrhina L.

Stem: annual; 1′ to 2′ tall; often branched; short hairs lying close to the stem; sometimes dark, sticky bands below the upper nodes.

Leaves: opposite; linear with rounded tips; 1″ by 1/8″ but variable in size; sessile; bulb-based hairs on the lower side.

Inflorescence: cymes at the stem tip and from the upper axils.

Flowers: petals pink to white, mostly hidden by the inflated calyx, 1/4″ long; petals protrude only 1/16″ to 1/8″; flowering from late May to mid-June.

Fruits: somewhat elongated capsules, 1/4″ in diameter within the calyx; fruiting begins in mid-June.

Habitat: on dry, gravelly, or sandy prairies and in other open places, often where some disturbance has occurred.

CISTACEAE
Rockrose Family

Frost weed
Helianthemum bicknellii Fern.

Stem: perennial; 6″ to 1 1/2′ tall; numerous short branches toward the top; hairy; lower leaves often drop, upper leaves very crowded.

Leaves: alternate; oblong, widest above the middle, tapering bases and rounded tips; lower, 1″ by 1/4″, upper, 1/2″ by 1/8″; sessile; short hairs above and below.

Inflorescence: first flowers in a terminal raceme and with petals, later flowers in clusters in the leaf axils and without petals.

Flowers: petals yellow, 1/2″ long, triangular; sepals hairy; many stamens; flowering from mid-June to early July; later flowers develop without petals.

Fruits: tiny capsules 1/16″ in diameter, within the calyx; fruiting begins in early July.

Habitat: frequent on sandy or rocky prairies.

Another frost weed, *Helianthemum canadense* (L.) Michx., barely enters Iowa. It is very similar to *H. bicknellii* except the outer two sepals are shorter than the inner three on the flowers bearing petals. In *H. bicknellii*, all the sepals are the same height. *H. canadense* is rare in northeast Iowa.

EUPHORBIACEAE
Spurge Family

Flowering spurge
Euphorbia corollata L.

Stem: perennial; 2′ to 3′ tall; unbranched below the inflorescence; smooth.

Leaves: alternate below, opposite in the inflorescence; oval with rounded bases and tips, sessile; 1 1/2″ by 1/2″; smooth margins; smooth above and below.

Inflorescence: becoming much branched to 1′ tall by 1 1/2′ wide at maturity; branching repeatedly from below opening flowers (cyme).

Flowers: "flower" is a cup (1/16″ wide) with five white petal-like bracts (3/32″ long) on the margin, within the cup are several male flowers, each with a single stamen, and a stalked female flower consisting of a single pistil with a three-lobed ovary; flowering from early July to late August.

Fruits: seedpods are three-lobed capsules 3/16″ in diameter on a curved stalk, the capsule thus hanging over the edge of the "flower cup"; each lobe contains one seed; seeds are "shot" from the capsule upon maturity; fruiting begins in early August.

Habitat: common on dry and sandy prairies; also on mesic prairies and roadsides.

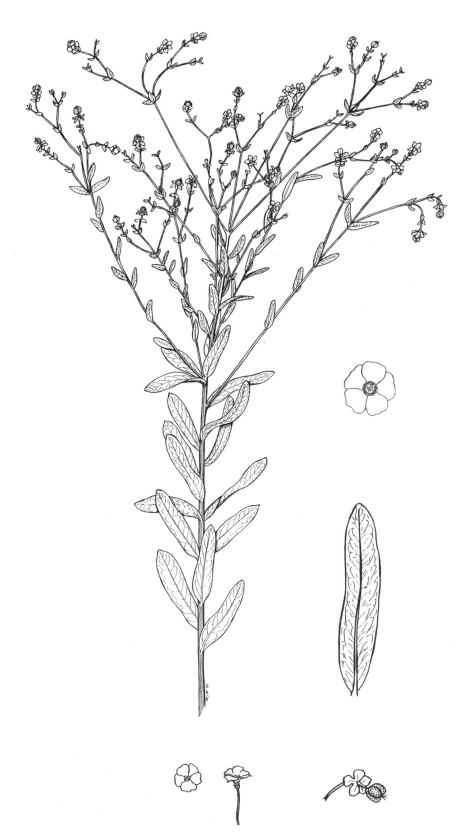

Toothed spurge
Euphorbia dentata Michx.

Euphorbia dentata is similar to
E. corollata except *branching occurs
all along the stem*. The stem is *sparsely
hairy*, and the leaves are *opposite* and
larger (1 1/2″ by 5/8″ to 2 1/2″ by 5/8″).
The leaves are on *short leaf stalks*, are
more pointed at the tip, and *have
toothed margins toward the tips*. The
inflorescence is *much smaller, with
clusters of "flowers" on short opposite
branches*. The "flowers" have *green-
fringed margins*, with similar stalked
ovaries and seedpods. Flowering is
from mid-July to mid-September, and
fruiting begins in late July. The habitat
is *dry, gravelly*, and *sandy disturbed
places* such as roadsides, railroad
embankments, disturbed prairies,
and open woods.

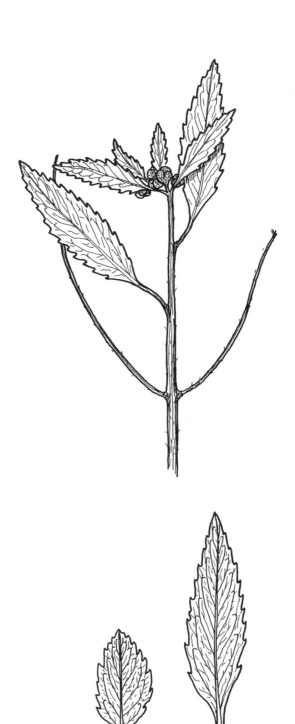

Spurge

Euphorbia glyptosperma Engelm.

Euphorbia glyptosperma is a very short plant (to 6″ tall), *much branched*, and *sprawling*. The leaves are *opposite, small* (3/8″ by 1/8″), and *densely packed on the stem*. The inflorescence is similar to *E. dentata*. The "flowers" have *white, reflexed margins* which are *sometimes fringed*. The ovaries and fruits are similar to *E. corollata*. Flowering is from early July to late August, and fruiting begins in mid-July. The habitat is *dry sandy roadsides and prairies and Loess Hills prairies*, usually with some disturbance.

Snow-on-the-mountain

Euphorbia marginata Pursh

Euphorbia marginata has upright stems, 1′ to 2′ tall and with few hairs. The leaves are alternate to opposite, *oval with rounded bases and pointed tips*, 2 1/2″ by 1 1/4″, and smooth above and below. Typically there are *three clusters of "flowers"* from the stem tip and upper leaf axils. The leaves below the flower stalks have *white margins*. The "flowers" are similar to *E. corollata* except the *"petals" are shorter*, and *each has a yellowish gland in the center*. The fruits are similar to *E. corollata*. Flowering is from mid-July to late August, and fruiting begins in early August. *E. marginata* is frequent on *dry prairies and Loess Hills prairies in western Iowa* and very infrequent elsewhere.

E. glyptosperma

E. glyptosperma

E. marginata

FABACEAE
Legume Family

Lead plant
Amorpha canescens Pursh

Stem: perennial shrub; 1′ to 2 1/2′ tall; branched; very hairy.

Leaves: alternate; pinnately divided; blades 2 1/2″ by 3/4″; twelve to twenty pairs of leaflets; leaflets oval with small points at the tips; very hairy above and below.

Inflorescence: racemes at the ends of the branches; 1″ to 4″ long; flowers densely packed.

Flowers: corolla purple, narrow, 3/8″ long, with yellow stamens projecting beyond the petals; calyx 3/16″ long, hairy; flowering from late June to mid-July.

Fruits: one-seeded pod 3/16″ long, densely covered with short hairs, curved style still attached; partially within calyx; fruiting begins in early July; fruits begin dropping in late July.

Habitat: frequent on dry and mesic prairies; also in woodland edges.

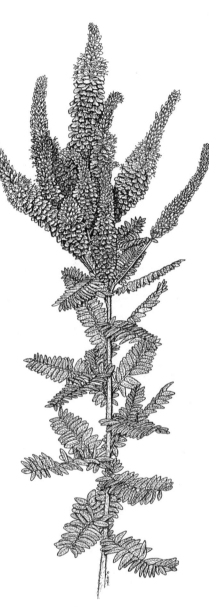

A. canescens

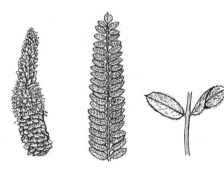

A. canescens

Fragrant false indigo
Amorpha nana Nutt.

Amorpha nana is similar to *A. canescens* but *not densely hairy* and with *black dots on the stems*. The *leaves are shorter and wider* (1 3/4″ by 1″), and there are fewer leaflets (*ten to fourteen pairs of leaflets*) per leaf with *black dots on the underside* of the leaflets. The inflorescence and flowers are similar except the *calyx has black dots*. The pods are similar in length but are more pointed. Flowering and fruiting dates are similar to *A. canescens*. *A. nana* is rare on dry, upland prairies.

A. nana

Milk vetch

Astragalus canadensis L.

Stem: perennial; 2′ to 3′ tall; branched
above; lightly hairy.

Leaves: alternate; pinnately compound; 5″ by 2 1/2″; leaflets oval with
rounded bases and tips, small points
at the tips, 1 1/4″ by 1/4″; hairy above
and below.

Inflorescence: raceme of crowded
flowers to 10″ long on leafless, hairy
stalks from upper leaf axils.

Flowers: corolla greenish cream;
showing 1/4″ to 1/2″ beyond calyx;
calyx tubular, 1/4″ long with points
1/16″ long, hairy; flowering from
mid-July to early August.

Fruits: pod 1/2″ long, smooth; fruiting
begins in late July.

Habitat: frequent on wet to dry prairies and in open woods, on roadsides, and in open places; more
common with some disturbance.

Ground plum
Astragalus crassicarpus Nutt.

Stem: perennial; sprawling, sometimes 2′ tall; often multistemmed from a common root crown; sparsely hairy.

Leaves: alternate; pinnately compound; 2 1/2″ by 3/4″; basal leaflets (stipules) round, 1/8″ long; leaflets oval, with rounded bases and tips; 3/8″ by 1/8″; hairy above and below.

Inflorescence: racemes of crowded flowers, 3″ long; from the lower leaf axils; stem growth continues leaving the flowers and fruits near the ground.

Flowers: corolla purple (to whitish); protruding 1/2″ beyond calyx; calyx 1/4″ long, with dark hairs; flowers from early to late May.

Fruits: roundish pod, 3/4″ long by 1/2″ in diameter, smooth; tan and wrinkled at maturity; central membrane divides the pod into two cavities; seeds black, 1/16″ long; fruiting begins in late May.

Habitat: frequent on dry to gravelly prairies and Loess Hills prairies.

Seedpods do not seem to have predators or dispersal mechanisms. Often fruits from several years' production can be found beneath a plant.

Milk vetch
Astragalus agrestis Douglas ex D. Don

Astragalus agrestis is *much smaller* than *A. crassicarpus*, about 6″ tall and branching with *hairy stems*. The leaves are similar. In the inflorescence the *racemes are shorter* (1″ long). The flowers are slightly larger, and the fruits are *smaller, with elongate pods* (3/8″ by 1/8″) *covered with long, white hairs*. Flowering is from mid-May to mid-June, and fruiting begins in late May. *A. agrestis* is very uncommon on dry prairies.

A. lotiflorus

Some authorities recognize a second species, *A. goniatus* Nutt., on the basis of larger size, larger flowers, later flowering, and less hairy pods.

A. agrestis

Milk vetch
Astragalus lotiflorus Hooker

Astragalus lotiflorus is *much smaller* than *A. crassicarpus* with short *sprawling* stems (3″ long) that are hairy and branching at ground level from a large rootstock. The leaves are similar with *slightly smaller leaflets* (1/2″ by 3/16″). The inflorescence is similar with slightly smaller *pale red (to white)* flowers. The calyx is *very hairy. The fruits are elongate* (3/4″ by 3/16″), *inflated, hairy*, and darkening to reddish brown to tan at maturity. Flowering is from early to late May, and fruiting begins in late May. *A. lotiflorus* is frequent on dry, gravelly, and Loess Hills prairies.

Bent milk vetch, *A. distortus* T. & G., with tiny leaflets, purple flowers, and smaller fruits, but similar in general growth pattern to *A. lotiflorus*, is very infrequent on dry and sandy habitats.

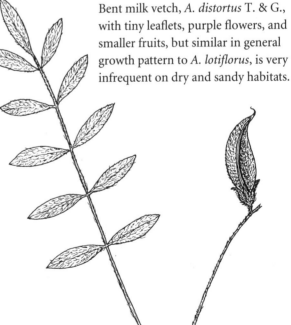

A. lotiflorus

False indigo

Cream false indigo
Baptisia bracteata Muhl. ex Ell.
   B. leucophaea Nutt.

Stem: perennial; 2′ tall; widely
   branched, shrublike; very hairy,
   yellowish in the spring.

Leaves: alternate; compound leaves
   1 5/8″ by 2″, nearly sessile; three
   leaflets, the middle one 1 5/8″ by
   1/2″, wider above the middle, blunt-
   pointed tip; large basal leaflets (stipules); very hairy.

Inflorescence: horizontal racemes or
   drooping at the sides of plant, 6″
   to 12″ long; each flower on a hairy
   stalk, about 1″ long.

Flowers: corolla pale yellow to cream,
   1″ long; calyx 5/16″ long, wide-
   spreading, blunt tips, hairy; flowering from mid-May to early June.

Fruits: inflated pods are 1 1/2″ long by
   1/2″ in diameter with a 1/4″ beak;
   black at maturity; seeds are about
   1/8″ long; fruiting begins in late May.

Habitat: frequent on dry to moist
   prairies and in open woods in
   northeastern Iowa.

Weevils and other seed predators often
destroy the seeds before the seeds
mature. In the fall the stems and leaves
become blackened and detach from
the rootstock, the plant becoming a
"tumbleweed."

White wild indigo
White false indigo
Baptisia lactea (Raf.) Thieret
B. *leucantha* T. & G.

Stem: perennial; 2 1/2′ to 3′ tall with
a flower stalk to 6′; branching to
become shrublike, smooth.

Leaves: alternate; compound leaves
1 1/2″ by 2″ on 1/2″ leaf stalks; three
leaflets, the middle one 1 1/2″ by
5/8″, nearly oval; smooth above and
below; turning black at the end of
the season.

Inflorescence: one or two racemes
2′ or more long with numerous
flowers about 1/2″ apart; projecting
above the leaves.

Flowers: corolla white to cream,
3/4″ to 7/8″ long on 1/2″ stalks; calyx
1/4″ to 3/8″ long, wide-spreading
tips produce a cup that persists into
fruiting; flowering from mid-June
to mid-July.

Fruits: inflated pods, 1″ long by 1/2″
in diameter; smooth; turning black
at maturity; fruiting begins in late
June.

Habitat: frequent on low to upland
prairies; sometimes found on roadsides and other open sites.

Weevils often eat the seeds in the pods
before the seeds mature.

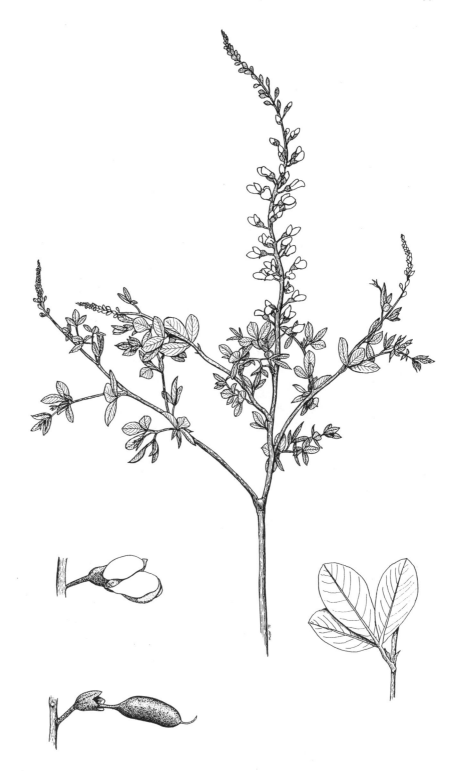

Partridge pea
Chamaecrista fasciculata (Michx.)
Greene
   Cassia fasciculata Michx.

Stem: annual; 1 1/2′ to 3′ tall; unbranched; hairy.

Leaves: alternate; pinnately compound, 2″ by 3/4″; leaflets 1/2″ by 1/8″, rounded bases, tips rounded with a point, eight to twelve pairs per leaf; leaflets with stiff marginal hairs.

Inflorescence: one or a few flowers in the leaf axils on the upper half of the stem.

Flowers: petals yellow, 1/2″ long, the flower cup-shaped; brown stamens, 3/16″ long; sepals lance-shaped, 1/4″ long; flowering from mid-July to mid-August.

Fruits: pods, 2″ by 3/16″, flat, brown, with hairy margins; seeds squarish, flat, 1/8″ long, black, shiny; as they dry pods split into two halves and twist, scattering the seeds; fruiting begins in late July; seeds are shed beginning in mid-August.

Habitat: common on dry, sandy prairies, on roadsides, and in open places.

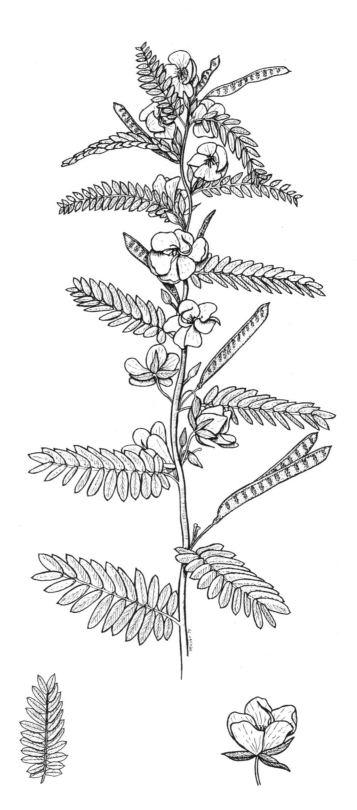

Rattle box
Crotalaria sagittalis L.

Stem: annual; 1′ tall; branching near the base; stiff-hairy.

Leaves: alternate; oval with tapered bases and pointed tips; sessile; 1 5/8″ by 7/16″; basal leaflets (stipules) pointed, stiff-hairy, extending on to the stem as a wing; stiff-hairy above and below.

Inflorescence: one to several flowers on 1″ to 4″ hairy stalks from the upper leaf axils.

Flowers: corolla yellow, not extending much beyond the calyx; calyx 3/8″ long, deeply lobed with sharp points, hairy; flowering over a long period from mid-July to late September.

Fruits: pods inflated, 3/4″ long by 3/8″ in diameter, light tan becoming dark brown at maturity; seeds rattle within when shaken, fruiting begins in late July.

Habitat: infrequent on sandy prairies, also in open woods, in ridge openings, and on sand dunes; more common in the southern part of Iowa.

Dalea
Dalea enneandra Nutt.

Stem: perennial; 2′ to 3′ tall; branching widely above the middle; smooth.

Leaves: alternate; pinnately compound, 1/2″ by 5/16″; usually five or more leaflets, very narrow (3/16″ by 1/32″), black dots on the leaflets and midrib; leaves crowded on robust specimens.

Inflorescence: few-flowered spikes at the ends of the branches, two or three (up to twelve) flowers; spike 1/2″ to 2″ long; each flower attached just above a leafy bract, 1/8″ long with black dots.

Flowers: corolla white, about 3/8″ long, petals spreading; calyx 1/4″ long, with pointed tips, very hairy; flowering from mid-July to mid-August.

Fruits: pod matures within the calyx and does not open; fruiting begins in early August.

Habitat: common on the west-facing Loess Hills prairies in western Iowa.

Foxtail dalea
Dalea leporina (Aiton) Bullock
   D. alopecuroides Willd.

Dalea leporina is *shorter* than *D. enneandra* (1′ to 2′ tall), *annual*, with branching on the lower part of the stem. The *leaves are larger* (2″ by 3/4″) with *larger leaflets* (3/8″ by 3/16″) and with black dots. The *flowers are crowded* in spikes at the ends of the branches and are 1/2″ to 1″ long. The corolla is white or pink, and the calyx is narrow and very hairy. Flowering is from late August to mid-September, and fruiting begins in late August. *D. leporina* is infrequent on moist, sandy prairies and in open places.

D. leporina

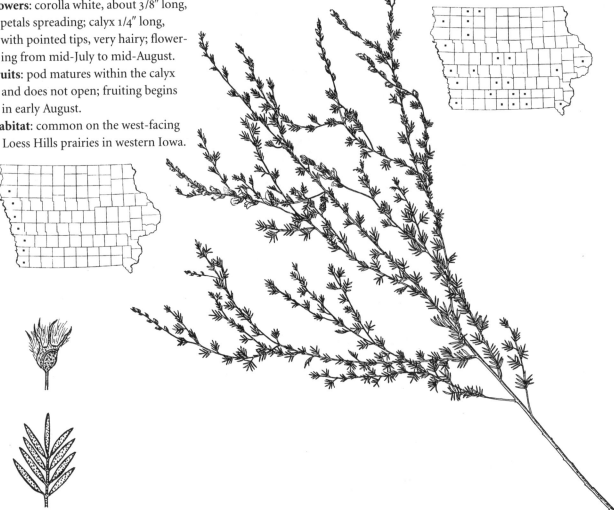

D. enneandra

D. enneandra

White prairie clover
Dalea candida Willd.
  Petalostemon candidum (Willd.)
  Michx.

Stem: perennial; 1′ to 3′ tall; sometimes branched toward the top;
  smooth.
Leaves: alternate; pinnately compound, 1 5/8″ by 1 1/4″; leaflets
  elongate-oval, three to four pairs
  of leaflets plus the tip leaflet; tip
  leaflet 5/8″ by 3/16″; smooth above
  and below but with black dots
  below.
Inflorescence: dense elongate heads at
  the stem tip and branches; head
  usually 1″ to 1 1/2″ long by 1/4″ in
  diameter; flowering from bottom
  to top.
Flowers: corolla white, to 1/4″ long;
  calyx 3/32″ long, with short, sharp
  tips, finely hairy; flowering from
  early July to early August; occasionally flowering into September.
Fruits: single-seeded pod develops
  within the calyx; 1/16″ in diameter;
  calyx developing black dots between
  each lobe; fruiting begins in mid-
  July.
Habitat: frequent on dry to moist
  prairies; also on sandy prairies.

Some authorities separate a more
branched, taller type found on the
Loess Hills prairies of western Iowa as
var. *oligophylla* (Torrey) Shinners (syn-
onym—*Petalostemon occidentale*
[Heller] Fern.).

Purple prairie clover
Dalea purpurea Vent.
  Petalostemon purpureum (Vent.)
  Rydb.

Stem: perennial; 1 1/2′ to 2 1/2′ tall;
  sometimes branching above the
  middle; often several stems from the
  same root crown; hairy.
Leaves: alternate; pinnately compound, 1″ by 3/4″; leaflets very narrow, 3/8″ by 1/16″; usually about five
  leaflets; smooth but with black dots
  on the lower leaf surface.
Inflorescence: dense elongate heads on
  flower stalks from the stem tip and
  the upper leaf axils; 1″ to 1 1/2″ long;
  flowering from bottom to top.
Flowers: corolla purple (3/16″ long)
  with yellow stamens protruding,
  calyx 3/32″ long, hairy; flowering
  from early July to early August.
Fruits: single-seeded pod, developing
  within the calyx, 1/16″ in diameter;
  fruiting begins in mid-July; fruits
  remain on the plant into September.
Habitat: frequent on dry, rocky, and
  sandy prairies, also on moist prairies and prairie openings in open
  woods.

D. purpurea

Silky prairie clover
Dalea villosa (Nutt.) Sprengel
  Petalostemon villosum Nutt.

Dalea villosa is similar to *D. purpurea*
in size. The leaves are similar but are
more crowded and more *hairy* and also
have *black dots*. The flowers are *light
purple*, and the *calyx is very hairy*.
Flowering and fruiting dates are similar. *D. villosa* grows on sandy prairies
and is *rare in Iowa*, known only from
Black Hawk County.

D. purpurea

Prairie mimosa

Illinois bundle flower
Desmanthus illinoensis (Michx.)
MacM.

Stem: perennial; 3′ to 4′ tall; somewhat branched; finely hairy on the
stem ridges.

Leaves: alternate; twice pinnately compound; 4″ by 3″; nearly sessile;
leaflets oval-elongate, 1/8″ by 1/32″;
stiff hairs on the margins.

Inflorescence: round heads on short
flower stalks at the tips of the
branches; up to twenty heads per
plant; heads 3/8″ in diameter.

Flowers: corolla white or greenish,
inconspicuous, 3/16″ long with protruding stamens; calyx 1/8″ long;
flowering from mid-July to early
August.

Fruits: flat curved pods, 1/2″ by 1/4″,
finely hairy, dark brown; clustered
in heads 1″ in diameter; fruiting
begins in late July.

Habitat: common on Loess Hills prairies and sandy soils in western Iowa,
very uncommon on sandy or rocky
prairies and waste places in the rest
of its range in the state.

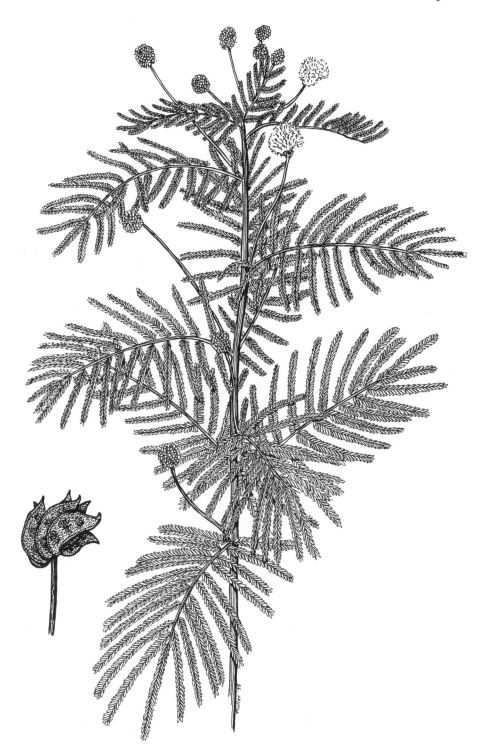

Showy tick trefoil
  Tick clover
  Desmodium canadense (L.) DC.

Stem: perennial; 3′ to 4′ tall;
  unbranched; hairy.

Leaves: alternate; divided into three
  leaflets, oval, with rounded bases
  and pointed tips, the center one
  2 1/2″ by 7/8″; 1/2″ leaf stalks; basal
  bracts (stipules) narrow, sharply
  pointed, 3/8″ long; hairy above and
  below.

Inflorescence: branching flower stalks
  from the stem tip and upper leaf
  axils; entire inflorescence 8″ to
  12″ long.

Flowers: corolla pink, 1/2″ long; calyx
  1/8″ long; flowering from mid-July
  to mid-August.

Fruits: pods, about 1″ long, flat, divided
  into about five rounded segments,
  one side more deeply indented than
  the other; surface of the pod covered
  with tiny hooked hairs by which the
  pod temporarily becomes attached
  to passing animals and thus is
  disseminated; fruiting begins in
  late July.

Habitat: frequent on dry to moist
  prairies, in open woods and ridge-
  top openings, on roadsides, and
  in other open places; often takes
  advantage of disturbed sites.

Illinois tick trefoil

Desmodium illinoense Gray

Desmodium illinoense is similar to *D. canadense* except it is *somewhat taller* (to 5′). The leaves are *slightly larger*, the *center leaflet stalk is longer*, and the *leaflets often have marginal hairs* in addition to *hairs on the upper and lower surfaces*. The basal bracts (stipules) are also *longer and wider*. The inflorescence is usually a *single raceme, 1′ to 2′ long*, with flowers *more widely spaced*. The flowers are very similar, *pink to white*, with fruits more *evenly lobed above and below*. Flowering is from mid-July to early August, and fruiting begins in late July. *D. illinoense* is frequent on dry to moist prairies and in open woods and is *more common in the southern half of Iowa*.

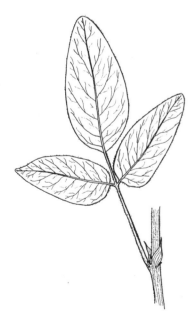

Wild licorice

Licorice-root

Glycyrrhiza lepidota Pursh

Stem: perennial; 2′ tall; little branched; sparsely hairy.

Leaves: alternate; pinnately compound, 4 1/2″ by 2″; about seven to eleven pairs of leaflets; leaflets 1 1/2″ by 1/2″, oval with somewhat rounded bases and tapered tips; smooth above and below.

Inflorescence: racemes at the ends of flower stalks from the stem tip and upper leaf axils; 1 1/2″ long; flower stalks hairy.

Flowers: corolla cream-colored, 1/2″ long; calyx 3/16″ long, deeply, sharply lobed; flowering from late June to mid-July.

Fruits: inflated pods, 1/2″ long by 1/8″ in diameter, brown; covered with hooked hairs, 3/32″ long; fruiting begins in early July.

Habitat: common on moist prairies and in other moist, open sites; less common toward the south and east in Iowa.

pod

Marsh vetchling
Lathyrus palustris L.

Stem: perennial; sprawling; to 2′ long; much branched; smooth.

Leaves: alternate; pinnately compound, 5″ by 3″; terminal three leaflets modified into tendrils; about five pairs of ordinary leaflets, oval, 1 1/2″ by 5/8″, with rounded bases and pointed tips, veins prominent below; basal bracts (stipules) pointed at both ends; smooth above and below.

Inflorescence: racemes on flower stalks from the upper leaf axils; flower stalks up to 6″ long, smooth.

Flowers: corolla purple to pink, about 5/8″ long; calyx 3/8″ long with pointed tips, smooth; flowering from late May to late June.

Fruits: pods cylindrical, 1 1/4″ long by 3/16″ in diameter, brown; at maturity the pod splits into halves which twist and scatter the seeds; fruiting begins in mid-June.

Habitat: infrequent on moist prairies and marshes.

unique stipule
on both species

Veiny pea
Wild pea
Lathyrus venosus Muhl. ex Willd.

Lathyrus venosus is similar to *L. palustris* except the *stem is somewhat hairy,* the leaves are slightly smaller, and the leaflets are more linear (1 1/4″ by 1/4″). The leaflets are *hairy below* with *less prominent veins* and *not strictly opposite each other.* There are *more flowers on each flower stalk.* Flowers and fruits are similar except the corolla is *pink to red* and the calyx is *hairy and red-purple.* Flowering is from early to mid-June, and fruiting begins in mid-June. *L. venosus* is infrequent on *upland prairies* and in *open woods.*

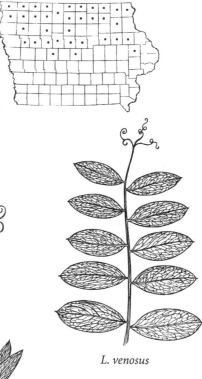

L. venosus

L. palustris

Round-headed bush clover
Lespedeza capitata Michx.

Stem: perennial; 2′ to 3′ tall; usually unbranched; stiff-hairy above; sometimes several stems from the same root crown.

Leaves: alternate; divided into three leaflets; middle leaflets 1″ by 3/8″; leaflets oval-elongate with tapering bases and pointed tips, hairs flattened on the lower surface, smooth above; leaves just below the inflorescence very hairy.

Inflorescence: heads at the stem tip and sessile in the upper leaf axils; the upper head the largest, about 1″ in diameter.

Flowers: corolla white, usually barely longer than the calyx; calyx 7/16″ long with sharp lobes; hairy, especially just before flowering; flowering from mid-August to early September.

Fruits: one-seeded pods develop within the calyx; 3/16″ long, flat and pointed at each end, brown, hairy; fruiting begins in late August.

Habitat: common on dry and sandy prairies; less common on moist prairies; also on dry and sandy roadsides and in open places.

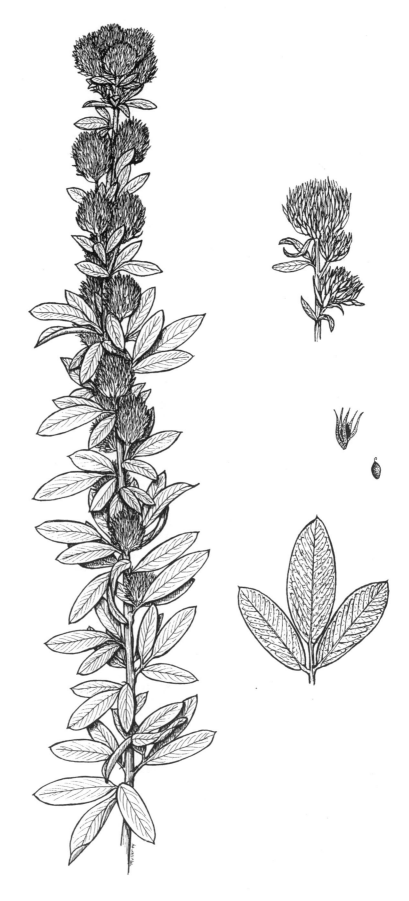

Prairie bush clover
Lespedeza leptostachya Engelm.

Lespedeza leptostachya is *shorter* than
L. capitata (2′ to 2 1/2′) with similar
leaves but *smaller, the middle leaflet 1″
by 1/4″*, and *hairy above and below*.
The inflorescence is a *raceme, 1 1/2″
to 2″ long* with *few flowers* and *not
crowded*. The flowers and fruits
are similar but *smaller. The calyx is
3/16″ long*, and *the pods are 1/8″ long*.
Flowering is from mid-August to
mid-September, and fruiting begins
in early September. *L. leptostachya* is
rare on dry to moist prairies. It is listed
as threatened both in Iowa and the
United States.

L. virginica

Slender bush clover
Lespedeza virginica (L.) Britton

Lespedeza virginica is similar to *L. capitata* except the *leaves are crowded*, and
the *leaflets are longer (1 3/8″ by 1/4″)*
and *more hairy above*. The flowers are
in *short racemes (1/2″ long)* from the
upper leaf axils. The corolla is *purple*
and *longer than the calyx*. The calyx
lobes *fall in fruit, exposing the pod*.
Flowering is from mid-August to mid-
September, and fruiting begins in mid-
September. *L. virginica* is frequent in
open, upland woods, in openings, and
on dry prairies.

L. leptostachya

Two other bush clovers, both non-
natives, might be encountered on dry
prairies in southern Iowa. Korean lespedeza, *L. stipulacea* Maxim., sometimes planted in grass-legume pasture
mixtures, occasionally escapes into
disturbed prairies and roadsides. It
is very short, 6″ to 12″ tall, and has
very small leaves and purple flowers
(3/16″ long). Silky bush clover, *L. cune-
ata* (Dumont-Cours.) G. Don, used as
a soil stabilizer, is about 3′ tall with
many branches, the leaves are very
crowded, and the flowers are 1/4″ long
with white petals marked with purple
spots. It also escapes into dry, disturbed habitats.

Locoweed

Oxytropis lambertii Pursh

Stem: perennial; very short with basal leaves.

Leaves: basal; pinnately compound, 5″ to 8″ by 2 1/2″; five to seven pairs of leaflets, 1 1/2″ by 3/16″, tapered to both ends; fine silky hairs above and below.

Inflorescence: up to fifteen flowers in 3″ to 4″ spikes on a hairy flower stalk that extends above the leaves; one to six flower stalks per plant.

Flowers: corolla purple, 3/4″ long; calyx hairy, purple, tubular, 3/8″ long with hair-tipped lobes; flowering from late May to mid-June.

Fruits: inflated pods, 3/4″ long by 1/4″ in diameter, long-tapering to the tips, hairy, dark brown; fruiting begins in early June.

Habitat: common on Loess Hills prairies and dry prairies.

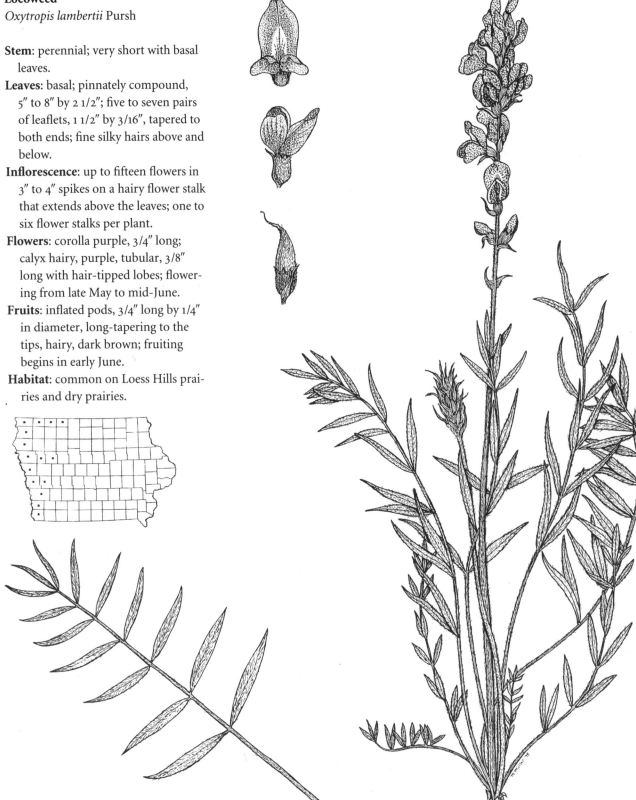

Prairie turnip
Pediomelum esculentum (Pursh) Rydb.
Psoralea esculenta Pursh

Stem: perennial; 6″ to 12″ tall; branched from the base; densely long-hairy.

Leaves: alternate; palmately compound with five leaflets, elongate-oval; the middle leaflet 1 1/2″ by 1/2″; leaf stalks 1 1/2″ to 2 1/2″ long; leaflets long-hairy below, smooth above.

Inflorescence: racemes at the ends of hairy, upward-curving flower stalks from the lower leaf axils, one to three per plant; racemes 1″ to 2 1/2″ long, the flowers crowded; 1/2″ long bracts below each flower, very hairy.

Flowers: corolla light blue, 3/4″ long; calyx 5/8″ long, hairy, with long, sharp lobes and stiff marginal hairs; flowering from mid-May to mid-June.

Fruits: one-seeded pod developing within the calyx; tapering to a persistent style 1/4″ long; fruiting begins in early June.

Habitat: common on dry, upland prairies to gravelly prairies and Loess Hills prairies in western Iowa.

Scurf-pea

Silverleaf scurf-pea

Pediomelum argophyllum (Pursh) Grimes

Psoralea argophylla Pursh

Stem: perennial; 1 1/2′ to 2 1/2′ tall; branching in the upper two-thirds of the stem; densely hairy.

Leaves: alternate; palmately compound; three to five leaflets, oval, 1″ by 1/2″; leaf stalks 1″ long; leaf surfaces white with dense hairs.

Inflorescence: clusters of several flowers at the nodes on 1″ to 3″ curving hairy flower stalks from the upper leaf axils; lower flowers blooming first.

Flowers: corolla dark blue, 1/4″ long, inconspicuous; calyx 1/8″ long, very hairy; flowering from mid-June to early July.

Fruits: one-seeded pod develops within the inflated calyx; fruiting begins in late June.

Habitat: frequent on upland prairies in the western half of Iowa, infrequent elsewhere; often grows in diffuse patches.

Scurfy pea

Psoralidium batesii Rydb.

Psoralea tenuiflora Pursh

Psoralidium batesii is *taller* than *P. argophyllum* (to 3′ tall), stands *more erect, is more branched*, and *is less hairy*. The *leaflets are narrower* (1″ by 3/16″) *and less hairy, with black spots above and below*. The inflorescence and flowers are similar except the calyx has *black spots*. The *pods are 3/16″ long and are also covered with black spots*. Flowering is from mid-June to early July, and fruiting begins in late June. *P. batesii* is infrequent on dry, rocky prairies and in woodland openings on ridges.

P. batesii

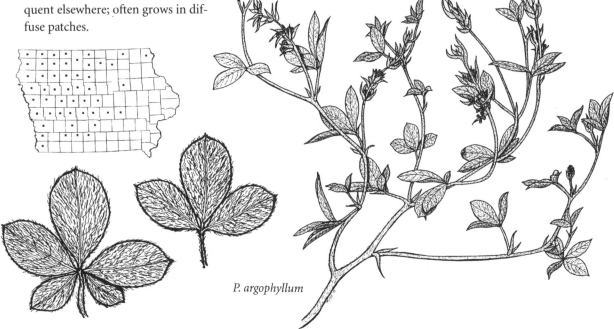

P. argophyllum

Wild bean

Trailing wild bean
Strophostyles helvula (L.) Ell.

Stem: annual; vine, up to 2′ long;
branching near the base; sparsely
long-hairy.

Leaves: alternate; pinnately compound, three to five leaflets; all but
the lowest having three-lobed
leaflets, 1 1/8″ by 7/8″; leaf stalks
about 3/4″ long; leaflets hairy above
and below.

Inflorescence: three- to five-flowered
racemes on 2″ flower stalks from the
leaf axils.

Flowers: corolla pink, 3/8″ long; calyx
1/8″ long, hairy; flowering from
mid-July to mid-September.

Fruits: pods long-tubular, 2″ long by
3/16″ in diameter, brown; splitting
and twisting when mature; usually
only one pod per flower stalk; seeds
squarish, 1/4″ long, dark brown;
fruiting begins in late July.

Habitat: frequent in sandy, rocky open
places but also on moist sites, usually with some disturbance; less frequent in northern Iowa.

S. leiosperma

Wild bean

Strophostyles leiosperma (T. & G.) Piper

Strophostyles leiosperma is similar to
S. helvula but the leaves have only *three
leaflets* which are *narrower* (1″ by 1/4″)
and *not lobed*. There is *one flower per
flower stalk* with a *bump on the stalk
below the flower*. The corolla is *shorter*
(1/4″), and the calyx is *densely hairy*.
Fruits are shorter (1″), finely hairy, and
less tubular. Flowering is from early
August to mid-September, and fruiting
begins in mid-August. *S. leiosperma* is
frequent in open, dry, rocky, and
sandy disturbed places to open woods.

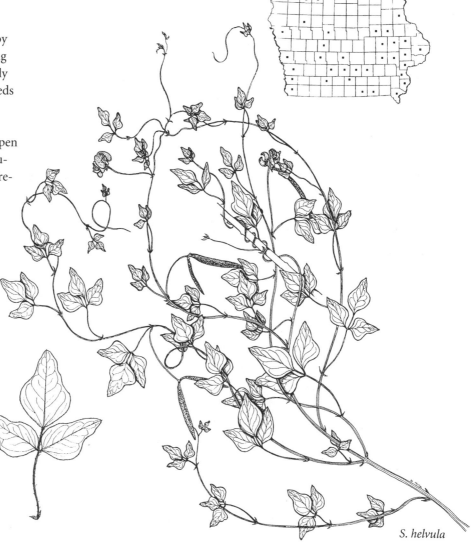

S. helvula

S. helvula

Goat's-rue
Tephrosia virginiana (L.) Pers.

Stem: perennial; 1′ to 2 1/2′ tall, not much branched above but sometimes with several branches from the base; hairy.

Leaves: alternate; 2″ to 4″ long; pinnately compound with twenty or more leaflets, each 1/2″ to 1″ by 3/32″ to 3/16″, with a tiny sharp tip; smooth above, long-hairy below.

Inflorescence: usually terminal, several-flowered raceme; about 2″ long.

Flowers: sweet-pea type flowers, large lower petals (standard) white with remaining petals red; calyx hairy with five equal points, 1/4″ long; flowering from early to late June.

Fruits: pods; long-tubular, 1 1/2″ to 2″ long, twisting and splitting open upon drying; seeds round, 1/8″ in diameter, flattened; fruiting begins in mid-June.

Habitat: infrequent on sandy prairies and dunes to sandy open woods.

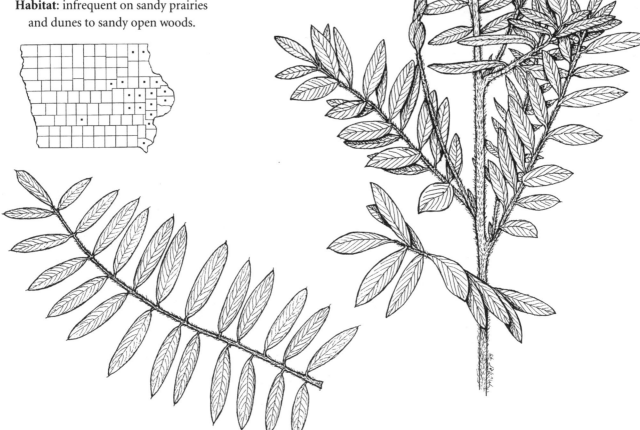

Vetch

  Purple vetch

Vicia americana Muhl. ex Willd.

Stem: perennial; sprawling to 2′ long;
  some branching; smooth.

Leaves: alternate; pinnately compound, variable in size, usually
  about 2 1/2″ by 1 1/4″; usually five
  pairs of leaflets and three tendrils at
  the leaf tip; leaflets 5/8″ by 1/4″, oval-
  elongate, smooth above and below;
  basal leaflets (stipules) 1/4″ long
  with several pointed lobes.

Inflorescence: racemes, five to seven
  flowered, on 2″ flower stalks from
  the upper axils.

Flowers: corolla purple, 5/8″ long;
  calyx tubular, 1/4″ long, smooth;
  flowering from mid-May to mid-
  June.

Fruits: pods long-tubular, 1 1/8″ long
  by 1/4″ in diameter, light brown,
  smooth; splitting and twisting when
  mature; fruiting begins in late May.

Habitat: common on moist and mesic
  prairies, also in moist, open woods.

On drier sites in far western Iowa, the
variety *minor* Hooker with narrow,
densely hairy leaflets is very infrequent. Two other vetches, both introduced, might occasionally be seen in
prairielike habitats. *Vicia villosa* Roth
is sometimes seeded on roadsides. It is
larger, with more leaflets on each leaf,
and with many flowers (fifteen to
twenty) on each raceme. The entire
plant is hairy. Common vetch, *V. sativa*
L., also escapes into roadsides and disturbed places. It has stems and leaves
similar to *V. americana*, but its flowers
are in pairs in the leaf axils.

GENTIANACEAE
Gentian Family

Pale gentian
Yellow gentian
Gentiana alba Muhl.
   G. flavida A. Gray

Stem: perennial; 2′ to 2 1/2′ tall; unbranched; smooth.

Leaves: opposite; oval with rounded, clasping bases and pointed tips; smooth edges; smooth above and below.

Inflorescence: clusters of several flowers in the uppermost leaf axils.

Flowers: corolla light yellow to greenish white, 1 5/8″ long with 3/16″ lobes; calyx 5/8″ long, lobes 1/4″, sharp-pointed; flowering from late August to mid-September.

Fruits: seedpods mature within the drying corollas; 1 1/2″ long, splitting into two halves; seeds numerous and tiny; fruiting begins in early September.

Habitat: very uncommon on prairies and in open places to open woods; moist to dry soil.

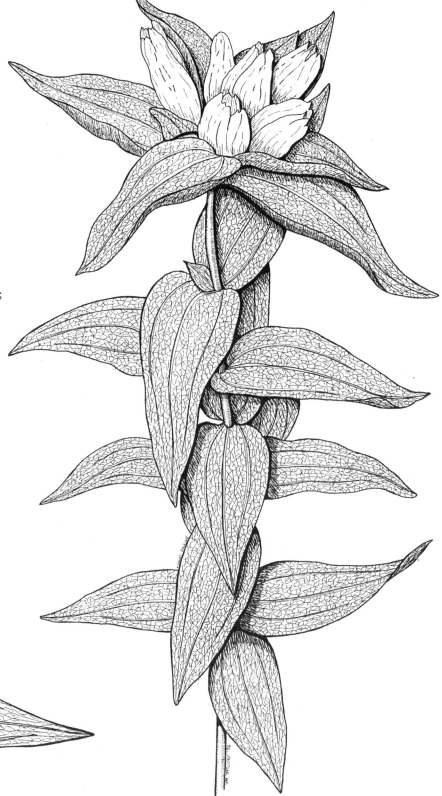

Downy gentian
Gentiana puberulenta J. Pringle
   G. puberula Michx.

Stem: perennial; 1′ to 1 1/2′ tall;
   unbranched; short-hairy; leaf bases
   continue down stem as hairy lines;
   leaves crowded, about 1/2″ apart.

Leaves: opposite; linear with rounded,
   clasping bases and pointed tips;
   1 1/2″ by 1/2″; smooth above and
   below.

Inflorescence: clusters of two to four
   flowers in the upper leaf axils.

Flowers: purple, corolla tube 1 1/4″ long
   with 3/16″ points, points recurved;
   sepals 5/8″ tall with abrupt, sharp
   tips; flowering from late August to
   late September.

Fruits: seedpods develop within the
   drying corollas; about 1″ long; opening into two halves; seeds numerous
   and tiny; fruiting begins in mid-
   August.

Habitat: very infrequent on mesic to
   dry prairies.

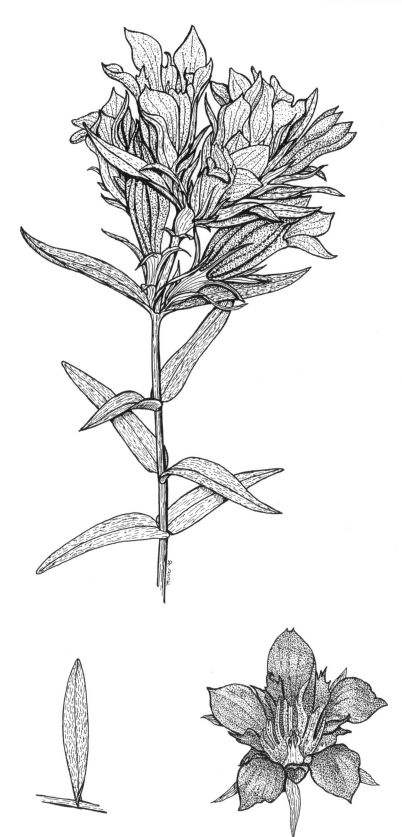

Bottle gentian

Gentiana andrewsii Griseb.

Gentiana andrewsii is similar to
G. puberulenta except the *stem is
smooth*. Sometimes several stems arise
from the same root crown. The leaves
are *larger* (2 1/2″ by 5/8″ to 3 1/2″ by
1 1/2″). The inflorescence sometimes
has *more flowers per cluster*, and the
corolla is closed at the top and is *lighter
purple*. The fruits are similar. Flowering is from early to late September,
and fruiting begins in mid-September.
G. andrewsii is frequent on *moist to wet
prairies*.

G. andrewsii

G. andrewsii

Fringed gentian

Gentianopsis crinita (Froel.) Ma
    Gentiana crinita Froel.

Gentianopsis crinita is *biennial* and
more branched on the upper half to
two-thirds of the stem than *G. puberulenta*. The *stem is smooth*. The leaves
are *ovate*, with a nearly chordate base
with *single flowers at the tips of
branches*. Plants have from one or two
flowers to twenty-five or more. The
*corolla lobes are 1/2″ to 5/8″ long,
fringed*, and *wide-spreading*. The calyx
is 1″ long with sharp tips. The fruits are
similar. Flowering is from mid- to late
September, and fruiting begins in late
September. *G. crinita* is *rare in moist,
sandy soils and fen habitats*.

G. crinita

GERANIACEAE
Geranium Family

Wild geranium
Geranium maculatum L.

Stem: perennial; 1′ to 2′ tall;
unbranched; few spreading hairs.

Leaves: basal and with one pair of
stem leaves below the inflorescence;
leaves deeply palmately lobed with
five principal lobes; blades 4″ across,
leaf stalks 6″ to 10″ long, stem leaves
smaller and on shorter leaf stalks;
stiff hairs above and below.

Inflorescence: several flowers at the
ends of branches and stem tip.

Flowers: petals red-pink, 1/2″ to 3/4″
long, wide-spreading; five lobed
style, five prominent stamens; sepals
1/4″ long with marginal and terminal hairs; flowering from early to
late May.

Fruits: seedpods narrow and pointed,
3/8″ long; seeds are released when
the lower portion of each fruit segment splits and coils toward the tip
of the fruit; fruiting begins in late
May.

Habitat: frequent in moist to mesic
woods and prairies.

Cranesbill
Carolina crane's bill
Geranium carolinianum L.

Geranium carolinianum is an *annual*,
8″ to 18″ tall with a finely hairy stem.
The leaves are similar to *G. maculatum*
except they are *alternate on the stem,
smaller* (1 3/4″ across the blade), and
more divided. The flower stalks arise
from the *axils of the stem leaves* with
clusters of up to fifteen flowers per stalk.
The flowers and fruits are similar.
Flowering is from late May to mid-
June, with fruiting beginning in mid-
June. *G. carolinianum* is *very uncommon* on prairies and in open places
with *dry, sandy, and gravelly soils*.

G. carolinianum

G. maculatum

HYPERICACEAE
St. John's Wort Family

Spotted St. John's wort
Hypericum punctatum Lam.

Stem: perennial; 1 1/2′ to 2 1/2′ tall; unbranched; smooth.

Leaves: opposite; oval, rounded bases and tips; 1 1/2″ by 5/8″; sessile; smooth above and below; tiny, black dots above and below.

Inflorescence: corymbiform; clusters of flowers on branching flower stalks at tips of branches.

Flowers: petals yellow with tiny black dots; flowers 3/8″ across; many stamens; sepals 1/8″ long, oval, pointed, black-dotted; flowering from early July to mid-August.

Fruits: seedpods a nearly round capsule with the remains of the style as a point at the top; 1/8″ in diameter; covered with black dots; fruiting begins in mid-July.

Habitat: frequent on upland prairies, in open places and open woods; more abundant with some disturbance.

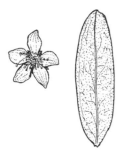

H. punctatum

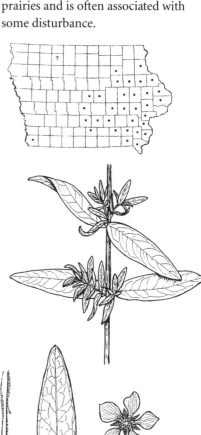

H. punctatum

Round-fruited St. John's wort
Hypericum sphaerocarpum Michx.

Hypericum sphaerocarpum is similar to *H. punctatum* except the stem is *slightly winged* and *somewhat shorter* (1′ to 2′). Short branches often develop from the stem leaf axils. The leaves are similar, up to 3″ by 5/8″, but *not black-dotted*. The inflorescence is similar. Flowers and fruits are similar but are *without black dots or spots*. Flowering is from late June to late July, with fruiting beginning in mid-July. *H. sphaerocarpum* is frequent in *moist to wet* open habitats, on roadsides, and on prairies and is often associated with some disturbance.

H. sphaerocarpum

LAMIACEAE
Mint Family

Water horehound
   Cut-leaved water-horehound
Lycopus americanus Muhl. ex Barton

Stem: perennial; 1′ to 2′ tall; sometimes branching at the middle or below; smooth.

Leaves: opposite; linear, tapering to both ends; sessile; 1 5/8″ by 3/8″ to 2″ by 5/8″; strongly toothed; smooth above and below.

Inflorescence: clusters of flowers at the nodes in the upper half to two-thirds of the plant.

Flowers: corolla white, 1/8″ long, two-lipped, the style exceeding the corolla; calyx 1/16″ tall with sharp tips; flowering from mid-July to mid-August.

Fruits: four nutlets, 1/16″ in diameter, develop within the calyx; fruiting begins in late July.

Habitat: common in wet open places such as riverbanks, sloughs, and low prairies.

variations

Wild bergamot

Horsemint

Monarda fistulosa L.

Stem: perennial; 2′ to 4′ tall;
unbranched; smooth.

Leaves: opposite; oval with rounded
bases and long-tapering sharp tips;
2 1/2″ by 1″; leaf stalks 1/2″ long;
margins sharply toothed; smooth
above and below.

Inflorescence: flowers grouped into
heads above leafy bracts at the stem
tips and the ends of branches from
the upper leaf axils.

Flowers: corolla purple, tubular,
5/8″ long, two-lipped, the upper
covering the stamens and style, the
lower recurved; calyx is narrow with
marginal teeth, 1/4″ long; flowering
from early July to early August.

Fruits: four nutlets, 1/32″ in diameter;
enclosed by the persistent calyx;
fruiting begins in late July.

Habitat: common on upland to moist
prairies; also on dry prairies, on
Loess Hills prairies, on roadsides,
and in open woods.

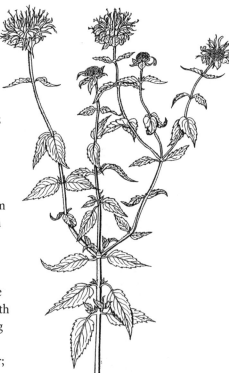

M. fistulosa

Upon ripening, nutlets ("seeds") read-
ily fall from the flowers when inverted.

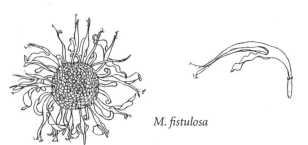

M. fistulosa

Spotted horsemint

Spotted bee balm, horsemint

Monarda punctata L.

Monarda punctata is similar to *M. fistulosa* except *shorter* (1 1/2′ to 2 1/2′ tall),
more branched above, and more *hairy*.
The *leaves are narrower* (2 1/4″ by 3/8″),
hairy above, and *very hairy below*, and
the *leaf surfaces have tiny clear dots*.
The flowers are *clustered at the nodes of
the upper branches*. The corolla is similar but *white to cream with purple spots*.
Fruits are similar. Flowering is from
mid-June to late July, and fruiting
begins in late June. *M. punctata* is
infrequent on prairies and stabilized
dunes *with very sandy soil* and *usually
with some disturbance*.

M. punctata

Self heal
Prunella vulgaris L.

Stem: perennial; 1′ to 2′ tall; unbranched below; smooth.

Leaves: opposite; oval with tapering bases, tapering to sharp tips; 3″ by 1 1/4″, and smaller; leaf stalks 1″ below to nearly absent above, hairy; margins nearly smooth; smooth above and below.

Inflorescence: flowers in a congested spike (to 1 1/2″ long) at the stem and branch tips; each flower nearly obscured by a leafy bract, kidney-shaped, with an abrupt sharp tip and stiff marginal hairs.

Flowers: purple corolla (1/2″ long), tubular, with two lips; calyx 1/4″ long with stiff-haired margins, elongating in fruit to become visible above the bract; flowering from early July to mid-August.

Fruits: four nutlets (1/16″ in diameter) maturing within the calyx; fruiting begins in mid-July.

Habitat: common in moist open woods, on roadsides and prairies; tolerant of disturbance.

Self heal is also found in Europe and Asia. There are both native and introduced varieties in Iowa. The introduced variety is found in more weedy habitats and has relatively wider leaves (see Eilers and Roosa 1994 and Gleason and Cronquist 1991).

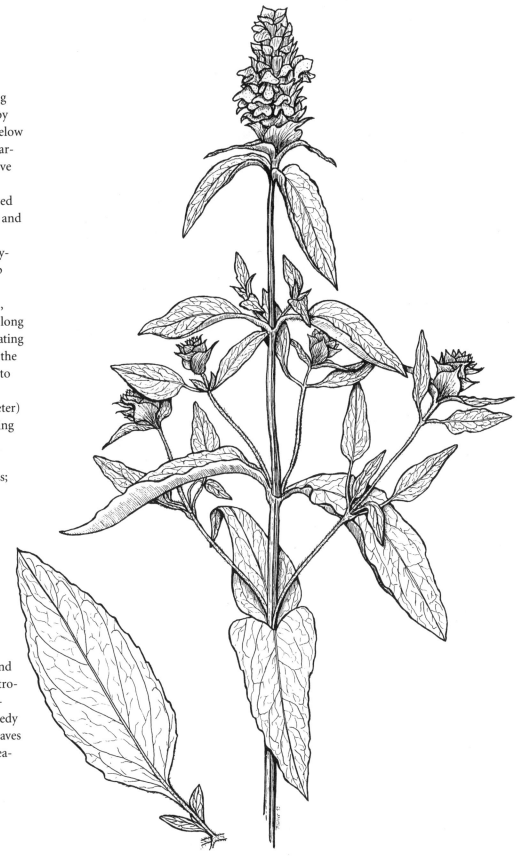

Common mountain mint
  Virginia mountain mint
  Pycnanthemum virginianum (L.)
  Dur. & Jackson

Stem: perennial; 2′ to 3′ tall;
  unbranched below; hairy on the
  stem angles, especially above.
Leaves: opposite; linear, long-tapering
  to pointed tips; sessile; 1 1/2″ by
  3/16″; with smooth margins; blades
  smooth above and below; tiny dots
  on the leaf surfaces.
Inflorescence: corymbiform cluster of
  heads from the stem tip and upper
  branches; each head 1/4″ across,
  underlain by hairy, overlapping
  bracts 1/8″ long.
Flowers: white corolla (1/4″ long) with
  purple spots inside, two-lipped;
  calyx 1/4″ tall, hairy; flowering from
  early July to mid-August.
Fruits: nutlets mature within the
  calyx, 1/32″ long; fruiting begins in
  mid-July.
Habitat: common on moist prairies,
  in marshes, and in other open
  places.

Slender mountain mint

Narrow-leaved mountain mint
Pycnanthemum tenuifolium Schrader

Pycnanthemum tenuifolium is similar
to *P. virginianum* except the stem is
smooth. The leaves are *short and
narrow* (1″ by 1/8″ and smaller) and
often with leafy shoots from the axils
of the stem leaves. The inflorescence is
similar, but the *bracts below the flower
heads are not hairy*. The flowers and
fruits are similar. Flowering is from
mid- to late July, and fruiting begins in
late July. *P. tenuifolium* is frequent on
upland prairies and in open places.

P. tenuifolium

Hairy mountain mint

Pycnanthemum pilosum Nutt.

Pycnanthemum pilosum is *taller* than
P. virginianum (to 4′ tall) and is *very
hairy on the stem angles and toward the
stem tip*. The leaves are similar but
wider (1/2″), with a *tapering base, and
are hairy below*. The inflorescence is
similar, but the flower stalks are *more
hairy* and the bracts beneath the heads
are *more elongate, hairy, and dotted*.
The flowers and fruits are similar.
Flowering is from late July to mid-
August, and fruiting begins in mid-
August. *P. pilosum* is frequent in
upland, open woods, on prairies, and
also on roadsides.

P. pilosum

Skullcap

Smaller skullcap

Scutellaria parvula Michx.

Stem: perennial; 4″ to 6″ tall; often branched; conspicuous stem angles with few hairs.

Leaves: opposite; oval with squarish bases and tapering to rounded tips; sessile; 1/2″ by 1/4″ and smaller; tiny hairs above and below.

Inflorescence: flowers in the axils of the upper leaves, two per node.

Flowers: corolla dark blue, 7/16″ long, two-lipped, hairy; calyx tubular, 5/16″ long, hairy on the veins, with a conspicuous enlargement on the upper surface, 3/16″ long; flowering from late May to mid-June.

Fruits: nutlets exposed when the upper part of the calyx drops off; fruiting begins in early June.

Habitat: frequent on dry prairies, in open woods, and in ridge openings; sometimes vigorous in disturbed areas; difficult to locate later in the season.

Germander

Wood sage

Teucrium canadense L.

Stem: perennial; 2′ to 3′ tall; branched
above; hairy on the stem angles.

Leaves: opposite; oval with short-
tapered bases, pointed tips; 3″ by 1″
and smaller; 1/2″ leaf stalks; toothed
margins; long-hairy above and
below.

Inflorescence: raceme, at the stem tip
and upper branches; to 6″ long at
maturity.

Flowers: corolla pink-purple, 3/4″ long,
tubular with long, lower lobe, upper
lobe missing; calyx funnel-shaped,
3/16″ long, hairy; flowering from
early July to mid-August.

Fruits: four nutlets develop within
calyx, 3/32″ long; calyx becomes
inflated in fruit; fruiting begins in
mid-July.

Habitat: common on moist prairies,
in open places, and in open woods;
also less common on drier sites;
somewhat weedy on disturbed sites.

Eilers and Roosa (1994) recognize two
varieties, var. *occidentale* (Gray) McCl.
& Epling and var. *virginicum* (L.)
Eaton.

LINACEAE
Flax Family

Wild flax
Yellow flax
Linum sulcatum Riddell

Stem: annual; 6″ to 2′ tall; grooved; sometimes branched above the middle; smooth.

Leaves: alternate; linear, 1/2″ by 1/16″ and smaller with sharp tips; sessile; smooth above and below; with stipular glands at the base of the leaf.

Inflorescence: branching, narrow to wide-spreading; from the stem tip and upper axils.

Flowers: corolla yellow, oval, blunt-tipped petals 1/4″ long; calyx 3/16″ long, with long, sharp lobes, covered with up-turning hairs and marginal hairs; flowering from early July to mid-August.

Fruits: seedpods are capsules; to 1/8″ in diameter, opening by numerous teeth, green-brown; seeds very small, 1/32″ long, oval, shiny, and slippery; fruiting begins in mid-July.

Habitat: frequent in western Iowa to very uncommon in eastern Iowa on dry prairies and Loess Hills prairies to sandy and upland prairies.

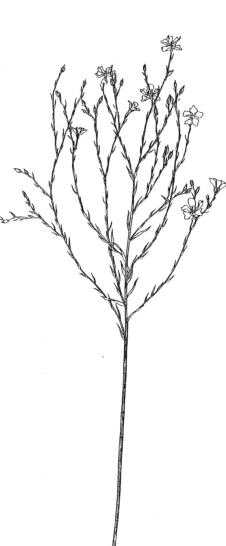

L. sulcatum

L. sulcatum

Stiff flax
Linum rigidum Pursh

Linum rigidum is about 1′ tall and is similar to *L. sulcatum*, but the stem is *not grooved*. The leaves are *somewhat narrower* (1/32″), with the *edges often rolling upward*. The *flowers are larger*, the petals *up to 1/2″ long*, and the five *style arms are enlarged at the tips*. The calyx is *divided to the base, hairy*, and *with very sharp tips*. The fruits split *into five sections. Flowering is earlier*, from *early June to mid-July*, and fruiting begins in late June. *L. rigidum* is infrequent on dry and Loess Hills prairies in western Iowa.

L. rigidum

LYTHRACEAE
Loosestrife Family

Winged loosestrife
Wing-angled loosestrife
Lythrum alatum Pursh

Stem: perennial; 2′ to 3′ tall; many ascending branches on the upper half; four-angled (winged); smooth.

Leaves: alternate; oval with rounded bases and pointed tips; 1 1/4″ by 3/8″ and smaller above; sessile; smooth above and below.

Inflorescence: single flowers in axils of leafy bracts.

Flowers: petals (6) blue to purple, 3/16″ long; sepals linear, 1/32″ long; both sepals and petals attached at the top of the narrow floral tube, 3/16″ long; flowering from late June to late July with some flowering continuing into late August.

Fruits: capsule developing within the floral tube; fruiting begins in early July.

Habitat: common in marshy, open places; also on wet prairies; less common in northern Iowa.

NYCTAGINACEAE
Four-o'clock Family

Wild four-o'clock
Mirabilis nyctaginea (Michx.) MacM.

Stem: perennial; 2′ to 4′ tall; branching
from below; lines of hairs between
the angles of the stem; becoming
woody later in the season.

Leaves: opposite; wide below, chordate, tapering to a pointed tip; 2 1/8″
by 1 1/8″ and larger; leaf stalks 1/2″
to 1″ to nearly sessile above; margins
uneven and undulating; smooth
above and below.

Inflorescence: crowded clusters of
flowers at the tips of short hairy
stalks; a pair of leaves below each
group of flower stalks.

Flowers: without petals; calyx purple,
resembling petals, tubular, 1/2″ long;
surrounded by five leafy bracts
(involucre) that resemble a calyx;
opening in late afternoon; flowering
from late May to mid-July.

Fruits: linear, 3/32″ long; fruiting
begins in mid-July.

Habitat: frequent on dry to mesic
prairies, also on roadsides and in
other open places; weedy and often
a pioneer on disturbed sites.

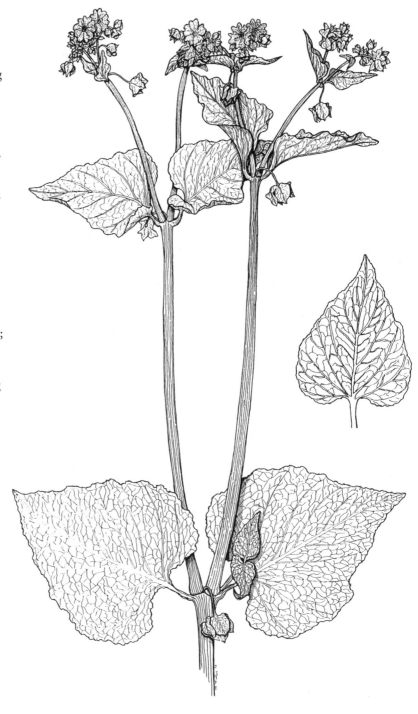

Hairy four-o'clock
Mirabilis hirsuta (Pursh) MacM.

Mirabilis hirsuta is often *taller* (4′) than *M. nyctaginea* and *hairy* on the upper stems. The *leaves taper* to each end and are *smaller* (3″ by 1/2″). There are two or three *much-branched flower stalks* per node on the upper stem, extending to 15″ in fruit. The involucre is similar, but the flowers are *smaller* (1/4″) and *pink* with projecting stamens. The fruit is 3/16″ long, hairy, and widest at the middle. It develops within the calyx. Flowering is from mid-July to early September, and fruiting begins in late August. *M. hirsuta* is frequent on dry and Loess Hills prairies, sandy prairies, and roadsides. It is often associated with disturbance.

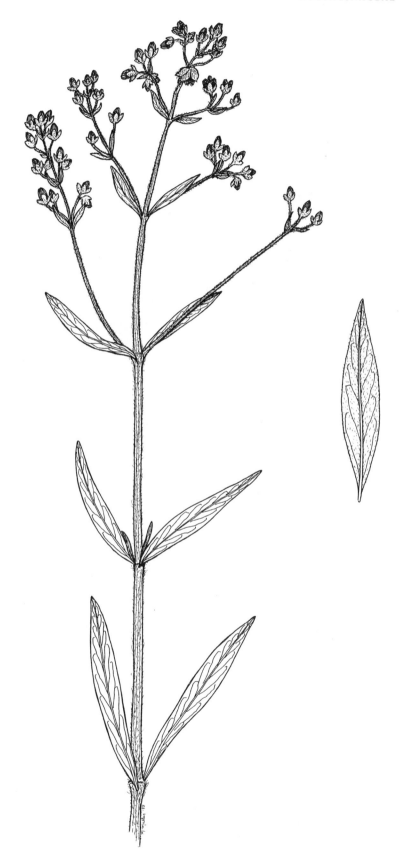

ONAGRACEAE
Evening Primrose Family

Toothed evening primrose
Calylophus serrulatus (Nutt.) Raven
  Oenothera serrulata Nutt.

Stem: perennial; 1′ to 1 1/2′ tall; usually
  unbranched; often several stems
  from the same root crown; hairy.

Leaves: alternate; oval, wider above the
  middle; 1 1/2″ by 3/16″ and smaller
  (especially in western Iowa); sessile;
  toothed margins (six to eight teeth
  per inch); hairy below, smooth
  above.

Inflorescence: spikes on leafy flower
  stalks from the stem tip and upper
  leaf axils.

Flowers: petals yellow, irregularly
  lobed, 1/4″ long on 1/4″ floral tube
  atop hairy ovary (inferior ovary),
  3/8″ long; sepals 3/16″ long, not
  reflexed; flowering from mid-June
  to mid-July.

Fruits: squarish capsule, 3/4″ long by
  3/32″ in diameter, hairy; fruiting
  begins in late June.

Habitat: frequent in western Iowa to
  very uncommon in eastern Iowa; on
  dry and rocky prairies to Loess Hills
  prairies.

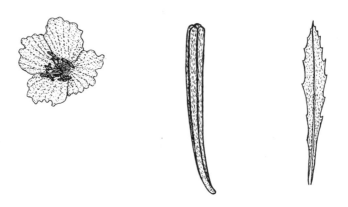

Prairie sundrops
Sundrops
Oenothera pilosella Raf.

Oenothera pilosella is similar to Caly-lophus serrulatus except it has longer hairs on the stems. The leaves are oval, tapering to both ends, larger (2 1/2″ by 1/2″), without teeth, and long-hairy above and below. The flowers are larger with petals 3/4″ long on a 7/8″ floral tube, the ovary is 1/2″ long, and the sepals are 1/2″ long and very hairy. The fruit is shorter (3/8″ long) with eight ridges and spreading hairs. Flowering is from early to late June, and fruiting begins in mid-June. *O. pilosella* is infrequent on *sandy, moist prairies.*

O. pilosella

O. perennis L., another sundrops, is a rare species and known only from a few locations in eastern Iowa. It is similar in size to *C. serrulatus*, but the leaves are *not toothed.* The fruit is *smaller* (3/8″ by 1/8″) and is *club-shaped, being larger in diameter beyond the middle,* and *with four winged ridges.* It is found on *wet prairies,* and it has been collected in the counties designated with P in the map above.

Ragged evening primrose
Cut-leaved evening primrose
Oenothera laciniata Hill

Oenothera laciniata is similar in height to Calylophus serrulatus but is more branched, more hairy, and annual. The leaves are larger (2 1/2″ by 1″) and deeply pinnately toothed (four to five teeth per side), with few, long hairs. The floral tube is longer (3/4″), but the sepals and petals are similar. *The fruits are similar in size but have spreading hairs, are without prominent ridges,* and usually are curved. Flowering is from early to late June, and fruiting begins in mid-June. *O. laciniata* is frequent in *sandy soils in open, disturbed places.*

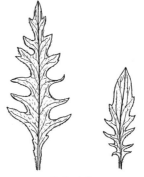

O. laciniata

Cinnamon willowherb

Purple-leaved willow-herb

Epilobium coloratum Biehler

Stem: perennial; 2′ to 3′ tall; much-branched above the middle; smooth except in the inflorescence.

Leaves: opposite; oval, tapering to both ends; 2 1/2″ by 3/8″; sessile; fine-toothed margins; smooth above and below.

Inflorescence: many short, branching flower stalks from the stem tip and upper leaf axils; hairy; to 12″ long, cylindrical.

Flowers: petals pink, 1/8″ long, sepals 1/16″ long, both attached at the top of the ovary (inferior ovary); ovary 3/4″ long, very narrow, 1/32″ in diameter; flowering from early August to early September.

Fruits: capsules, 1 1/2″ long by 1/16″ in diameter; splitting from the tips into four segments; seeds tiny, 1/32″ long, with hairy plumes attached at one end; fruiting begins in mid-August; seeds begin flying in early September.

Habitat: frequent in marshes, in low woods, on stream banks; often with some disturbance; seldom found on prairies; less frequent to the south and west.

Biennial gaura

Gaura biennis L.

   G. filiformis Small

Stem: biennial; 3′ to 4′ tall; branching from the middle; downy-hairy.

Leaves: alternate; oval with long tapering bases and pointed tips; 1 1/2″ by 7/16″; sessile; hairy above and below.

Inflorescence: spikes at ends of branches; with tiny leaves in the inflorescence; flowering from bottom to top with only a few flowers open at one time.

Flowers: petals white to red, 3/8″ long on a 3/16″ long floral tube above a hairy 1/8″ ovary; sepals 3/8″ long, curved backward, hairy; flowering from mid-June to mid-September.

Fruits: spindle-shaped capsule, with four prominent angles, 5/16″ long, hairy; fruiting begins in early July.

Habitat: infrequent on disturbed prairies and open sites, often in low-lying sandy soil.

A highly variable species with wide ranges in plant height, leaf size, flower size, and flowering time.

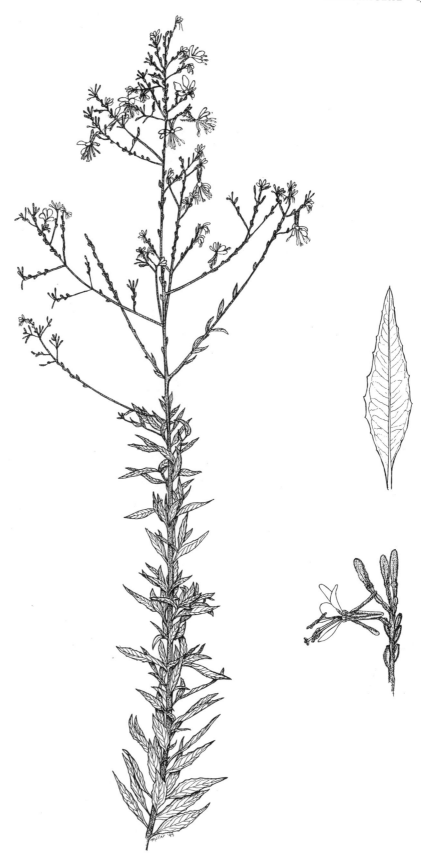

Scarlet gaura

Gaura coccinea Pursh

Stem: perennial; 1′ to 2′ tall; branched above; hairy.

Leaves: alternate; oval, 7/8″ by 1/4″; some with a large rounded tooth near the base on each side; upper leaves smaller; hairy above and below.

Inflorescence: spikes at the tip of leafless stalks from the upper branches; flowers very crowded when flowering but 1/4″ apart in fruit.

Flowers: petals red, 3/16″ long, attached to 3/16″ floral tube atop 1/8″ hairy ovary; sepals recurved, 3/16″ long; flowering from early June to mid-August.

Fruits: pear-shaped, four-angled capsules, 5/16″ long, hairy; above bracts of the same length; fruiting begins in mid-June.

Habitat: infrequent on dry Loess Hills prairies.

Gray evening primrose

Oenothera villosa Thunb.

   O. biennis L. var. *canescens* T. & G.

   O. strigosa (Rydb.) Mack. & Bush

Stem: biennial; 3′ to 6′ tall; branching above; hairy.

Leaves: alternate; elongate-oval, tapering to both ends; 5″ by 1″; sessile; margins weakly toothed; hairy above and below.

Inflorescence: spikes of flowers crowded at the ends of flower stalks from the stem tip and upper leaf axils.

Flowers: petals yellow, 7/8″ long on floral tubes 1 1/8″ long atop a 1/2″ ovary (inferior ovary); sepals 5/8″ long, hairy, recurved; flowering from mid-July to late August.

Fruits: elongate capsules, 1″ by 3/16″ in diameter, weakly angled, hairy, opening by four sections separating at the tip and recurving; fruiting begins in late July.

Habitat: frequent in disturbed places in mesic to moist soil on roadsides, on prairies, and in open woods.

O. villosa

O. villosa

Sand primrose

Oenothera rhombipetala Nutt. ex T. & G.

   O. clelandii D. Dietr. & Raven

Oenothera rhombipetala is *shorter* than *O. villosa* (2′ to 3′ tall). The *leaves are 2″ by 1/2″ or smaller*. The flowers are *smaller*, and the petals are *1/2″ long* and are *rhomboid-shaped*. Fruits are *1/2″ long*. Flowering is from early July to late September, and fruiting begins in mid-July. *O. rhombipetala* is frequent in *sandy soils* on prairies and in open places.

O. rhombipetala

OXALIDACEAE
Wood Sorrel Family

Violet wood sorrel
Oxalis violacea L.

Stem: perennial; buried, scaly, bulblike base.

Leaves: basal, compound with three leaflets at the end of smooth leaf stalks; the leaflets 1/2″ by 1/2″, widest above the middle, tapering to the base, and with a notch at the tip end (shamrock-shaped), the leaflets folding and drooping at night; the leaf stalks about 2″ long.

Inflorescence: umbel of flowers at the end of a flower stalk (about 4″ long), which surpasses the leaves.

Flowers: pale violet corolla, funnel-shaped, each petal 1/2″ long; calyx 5/32″ long, the lobes brown-tipped; flowering from late April to late May.

Fruits: capsules roundish, 5/16″ long by 1/4″ in diameter; remnants of the five style branches attached at the tips; drooping; fruiting begins in mid-May.

Habitat: frequent to common on dry, sandy, and Loess Hills prairies; also in dry, woodland openings, in open woods, and on moist prairies.

Flowers of two forms, pins and thrums. Pins have short stamens and long styles, while thrums have long stamens and short styles.

PLANTAGINACEAE
Plantain Family

Plantain
Plantago patagonica Jacq.
  P. purshii R. & S.

Stem: perennial; very short with basal
  leaves.

Leaves: basal; linear, 2″ to 4″ by 1/8″,
  pointed tips; sessile; long-hairy
  above, woolly below.

Inflorescence: raceme on a leafless
  flower stalk, 3″ to 6″ long; flowers
  crowded, raceme 1″ to 3″ long;
  woolly.

Flowers: four tiny white petals barely
  exceed the calyx; woolly calyx, 1/8″
  long; flowering from early June to
  early July.

Fruits: capsule develops within the
  calyx; smooth, lens-shaped; open by
  dropping of the top of the capsule;
  fruiting begins in mid-June.

Habitat: common on dry, sandy,
  rocky, and Loess Hills prairies in
  western Iowa; infrequent in eastern
  Iowa; also on roadsides, in pastures,
  and in other dry, disturbed places.

POLEMONIACEAE
Phlox Family

Prairie phlox
Phlox pilosa L.

Stem: perennial; 1′ to 2′ tall; sometimes branched above; hairy.

Leaves: opposite; lance-shaped; 1 1/2″ by 1/8″; sessile; hairy above and below.

Inflorescence: corymbiform to columnar; clusters of flowers on flower stalks from the stem tip and upper leaf axils.

Flowers: corolla commonly red but ranging from nearly white to lavender, the lighter variants with dark spots (nectar guides) at the base of the lobes of the corolla; floral tube 1/2″ long, the lobes 5/16″; calyx 1/2″ long with lance-shaped lobes, hairy; flowering from mid-May to early June; compared to central Iowa flowering in northern Iowa is delayed by three weeks and in southern Iowa by one week.

Fruits: round capsule, 1/8″ in diameter, light brown, breaking into three sections; seeds black, 1/16″ long, three per fruit, being thrown from the fruit as it opens explosively upon drying; fruiting begins in early June.

Habitat: common on mesic to dry prairies, also in open woods and woodland openings; occasionally on roadsides and in open, grassy places.

Wild phlox

Phlox maculata L.

Phlox maculata is similar to *P. pilosa* but is *slightly taller* (to 2 1/2′) with *purple blotches on the stem*. The leaves are *larger* (3″ by 1/2″), more or less *sheathing the stem*. The inflorescence is *cylindrical, up to 6″ long*. The corolla is similar in color and size, while the calyx is *shorter* (5/16″) with shorter lobes. The fruit is similar. Flowering is from *early to late June*, and fruiting begins in mid-June. *P. maculata* is frequent on *moist prairies and in marshes* in northeastern Iowa and is infrequent elsewhere within its range in the state.

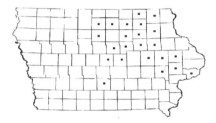

Cleft phlox, *Phlox bifida* Beck, is very uncommon in east-central Iowa. It is a *low-growing, sprawling* plant with leaves 5/8″ long. There are two or three flowers in clusters at the upper nodes, and each *petal is divided into two lobes*. *P. bifida* is rare in *sandy soil* on prairies and in open woods and is often associated with some disturbance.

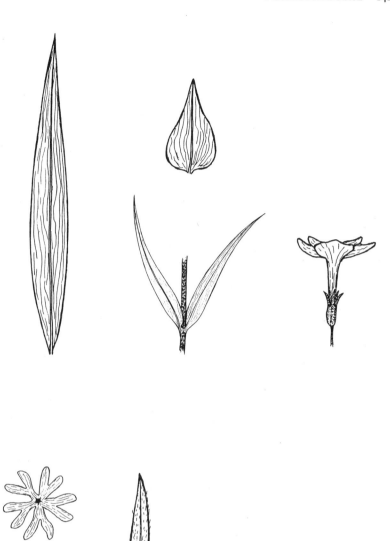

POLYGALACEAE
Milkwort Family

Field milkwort
Polygala sanguinea L.

Stem: annual; 8″ to 18″ tall; often several stems from one root crown; sometimes branched near the top; nearly hairless; leaf midribs continue down the stem as ridges.

Leaves: alternate; linear with tapering bases and blunt tips, 1 1/2″ (3/4″) by 1/8″; sessile; few hairs.

Inflorescence: raceme of tightly clustered flowers at the stem tip and ends of the branches; about 1/2″ long with new flowers being produced at the tips as the lower ones mature and drop off.

Flowers: pink corolla which fades to white as the flowers age, 3/16″ long; calyx 1/16″ long, sharp-pointed; flowering from early July to late August.

Fruits: two-seeded capsules that develop within the corollas; 1/16″ long; fruits drop off when mature, each leaving a tiny bract on the stem; fruiting begins in mid-July.

Habitat: frequent on moist to mesic prairies; also on dry and sandy prairies and in woodland openings.

Field milkwort is often overlooked because it is overtopped by the surrounding vegetation. Spreading the grasses near the ground may reveal this small plant.

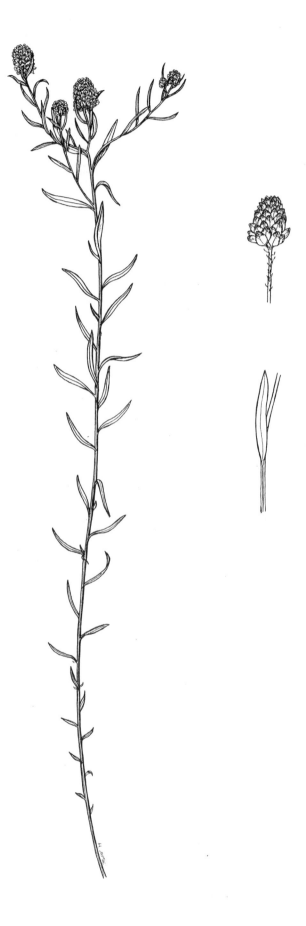

Seneca snakeroot

Polygala senega L.

Stem: perennial; about 1′ tall; sometimes several stems from one root
crown; finely hairy.

Leaves: alternate; oval with short-
tapered bases and long-tapered tips;
1 1/4″ by 1/4″, smaller toward the
base of the stem; sessile; finely serrated margins; slightly hairy below.

Inflorescence: raceme, 1/2″ to 1″ long;
flowering from bottom to top.

Flowers: corolla white, 1/8″ long; calyx
petal-like, 1/32″ long; flowering from
mid-May to late June.

Fruits: capsules, roundish, 1/8″ in
diameter, hairy; fruiting begins in
late May.

Habitat: infrequent in dry, open woods
and woodland openings to sandy
and upland prairies; most common
in northeast Iowa; becoming more
infrequent to the west and south.

Whorled milkwort
Polygala verticillata L.

Stem: annual; about 6″ tall; many
   branches; sprawling; smooth.
Leaves: whorled; usually five leaves,
   ranging from three to five, at each
   node; elongate, 1″ by 1/16″; sessile;
   smooth.
Inflorescence: raceme, 3/8″ long; on
   short flower stalks, from the upper
   nodes.
Flowers: corolla white, 1/16″ long;
   calyx of separate sepals, 1/32″ long,
   green with white margins; flowering
   from mid-July to late August.
Fruits: capsule develops within the
   corollas; 1/16″ long with prominent
   lines on the surface; fruiting begins
   in late July.
Habitat: infrequent on dry, sandy, and
   Loess Hills prairies; also in barren
   soil, moist swales, and open woods.

Some inflorescences flower over a
period of more than a month, producing new flowers at the tip while mature
fruits drop from the bottom. This species resembles *Galium* (bedstraw), but
the inflorescence and flowers are completely different, making identification
easier. Also, this milkwort is often
overlooked because of its small stature.

POLYGONACEAE
Smartweed Family

Pennsylvania smartweed
Polygonum pensylvanicum
  L. var. *laevigatum* Fern.

Stem: annual; 2′ to 4′ tall; branching
  above; reddish, smooth, with
  swollen nodes.

Leaves: alternate; oval, tapering to
  both ends; 3 1/2″ by 3/4″ to 4 1/2″
  by 1″; leaf stalks 1/4″ to 1/2″ long;
  sheaths surrounding the stems at the
  base of the leaf stalks 1/2″ long, with
  smooth margins; appressed-hairy
  on the margins, otherwise smooth.

Inflorescence: raceme; flowers closely
  packed; 1″ to 2″ long by 1/2″ in
  diameter; 3/16″ flower stalks stiff-
  hairy, from the stem tip and upper
  leaf axils.

Flowers: sepals pink, 5/16″ long, overlapping, the tip of the flower nearly
  closed; without petals; flowering
  from mid-July to mid-August.

Fruits: seedlike, flat, circular, shiny
  black, 1/8″ long, developing within
  the calyx; fruiting begins in early
  August; fruits begin dropping in
  September.

Habitat: common in moist, recently
  disturbed soil; on moist prairies, on
  roadsides, in cultivated fields, and in
  other open places.

PRIMULACEAE
Primrose Family

Shooting star
Dodecatheon meadia L.

Stem: perennial; basal leaves and smooth leafless flower stalk.

Leaves: basal; oval, widest above the middle, long-tapering to the leaf stalk; blades 6″ by 2″; margins smooth; smooth above and below.

Inflorescence: umbel of about a dozen flowers on a smooth, leafless flower stalk, 1′ to 1 1/2′ tall; flower buds erect, curving downward in flower, and again becoming erect in fruit.

Flowers: petals (5) pink (but varying from white to lavender), reflexed, 3/4″ long, the grouped stamens forming a point 5/16″ long, black nectaries between the bases of the stamens; calyx 3/8″ long with pointed lobes, hidden beneath the petals during flowering; flowering from early to late May.

Fruits: oval capsules, 1/2″ long, opening at the tips with five points; seeds, 1/32″ in diameter, brown, remaining in the upright fruits through the summer; fruiting begins in late May.

Habitat: frequent on moist to dry and sandy prairies, in woodland openings, and in open woods.

Some seeds often remain within the ripened capsule into the summer and can be easily harvested by tipping the capsule into a container and shaking the capsule.

Fringed loosestrife

Lysimachia ciliata L.

Stem: perennial; 1 1/2′ to 2 1/2′ tall; some branching above the middle; few hairs.

Leaves: opposite; oval with rounded bases and pointed tips; 3 1/4″ by 1 1/2″; leaf stalks 1/2″ with stiff hairs on the angles; smooth above and below.

Inflorescence: flowers on 1″ to 1 1/2″ flower stalks, from the upper leaf axils, two flowers per node; the flowers drooping and facing downward.

Flowers: corolla yellow, 1/2″ long, flat-faced, hairs near the base; small projections between the stamens; calyx 3/8″ long, sharp-pointed; flowering from late June to early August.

Fruits: oval capsules 3/16″ long, opening at the tips with five points; fruiting begins in early July.

Habitat: common on moist soils on prairies, in open woods, on roadsides, and in other open places; often associated with some disturbance.

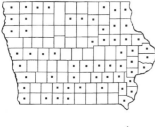

RANUNCULACEAE
Buttercup Family

Canada anemone
Anemone canadensis L.

Stem: perennial; 1′ to 1 1/2′ tall, somewhat sprawling; few long hairs.

Leaves: basal with leaf stalks up to 6″;
 blades deeply lobed into three broad
 parts, each shallowly lobed into
 threes; the blades 2 1/2″ by 4″; two
 leaves just below the inflorescence
 opposite, sessile, the blade 2″ by 3″;
 hairy above and below.

Inflorescence: a single flower per
 flower stalk; usually two stalks per
 plant.

Flowers: sepals petal-like, white,
 1/2″ long, usually five per flower
 (no petals); many pale yellow stamens; flowering from mid-May to
 late June.

Fruits: one-seeded, flattened, circular
 with a prominent wing, hairy; 1/8″
 long with a persistent style; the
 numerous fruits of each flower
 forming a ball 3/8″ in diameter;
 fruiting begins in mid-June, and
 fruits begin dropping in late June.

Habitat: common on moist prairies
 and in open woods; also on roadsides and in other open places; often
 forming large patches.

Windflower

Thimbleweed

Anemone cylindrica Gray

Anemone cylindrica is taller than *A. canadensis*, 1 1/2′ to 2 1/2′ tall, and branching near the middle where two opposite *stalked leaves* are attached. The stems are long-hairy. The leaves are *smaller* and *more deeply lobed*, with blades 1 1/2″ by 2 1/2″. Single flowers are borne *on long stalks* from the stem tips, one to four per plant, occasionally as many as six. Flowers are similar with white sepals *1/4″ long*. The *central mound of pistils is 1/4″ wide and tall.* The fruits are one-seeded, *1/16″ long*, and *covered with cottony hairs*. The *receptacle elongates in fruit to form a "thimble" 3/4″ to 1 1/4″ long by 1/4″ in diameter*. Flowering is from early June to early July, and fruiting begins in mid-June. Fruits begin to fly in mid-August, but some persist on the plant into winter. *A. cylindrica* is common on dry prairies and in open woods as well as on roadsides.

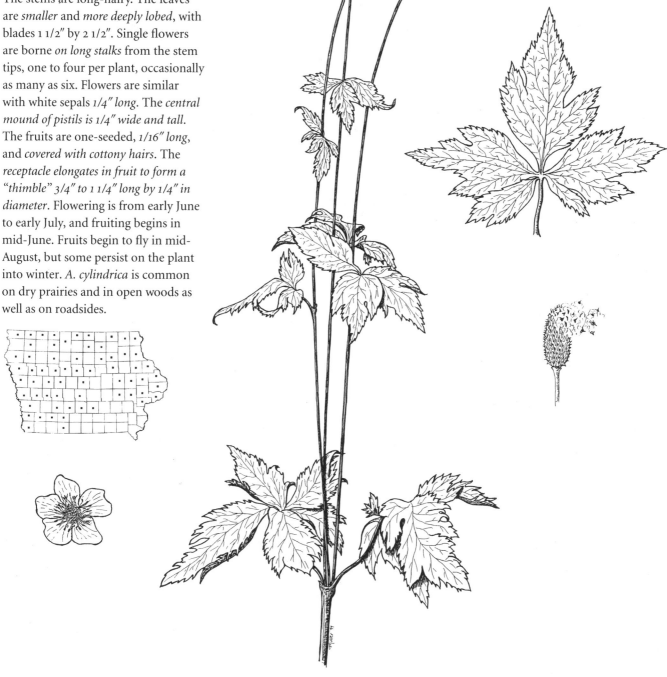

Prairie larkspur

Delphinium virescens Nutt.

Stem: perennial; 2′ to 3′ tall; seldom branched; short-hairy.

Leaves: alternate; deeply, narrowly lobed into threes; each lobe again divided; leaf stalks 3″ long, hairy; blades 2 1/2″ by 1 1/4″, hairy above and below.

Inflorescence: raceme with overlapping flowers; 6″ or longer.

Flowers: both calyx and corolla white; upper sepals elongated into a curved spur, 1/2″ long; lower petals 3/4″ long, cupped; flowering from early June to mid-July.

Fruits: three-lobed capsules, 5/8″ long by 1/4″ in diameter; tips of the lobes curved outward; often only the lower flowers setting fruit; fruiting begins in mid-June.

Habitat: common on dry and sandy prairies to upland prairies, especially in northwestern Iowa, to infrequent in southeast Iowa.

Pasque flower

Pulsatilla patens (L.) P. Miller
Anemone patens L.

Stem: perennial; basal leaves and a flower stalk with one pair of sessile leaves; densely long-hairy.

Leaves: basal with leaf stalks up to 6″; blades deeply lobed into three linear segments each again with three lobes; blades 3 1/2″ by 2 1/2″; stem leaves sessile, smaller, but with similar lobing; very hairy above and below.

Inflorescence: flower stalk 4″ to 6″ long, with a single flower; very hairy.

Flowers: sepals petal-like, blue or purple (sometimes almost white), 1/2″ to 1″ long (no petals); many yellow stamens; flowering from mid-April to mid-May.

Fruits: one-seeded, 1/16″ long with the elongate, hairy styles attached; styles to 1 1/4″ long; mature fruits make a hairy ball 2 1/2″ in diameter at top of flower stalk; fruiting begins in early May, and fruits begin to fall in late May.

Habitat: infrequent on dry, rocky prairies and in woodland openings.

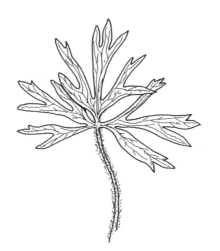

Early buttercup

Ranunculus fascicularis Muhl.

Stem: perennial; 6″ to 10″ tall; often
with several stems from the same
root crown; smooth.

Leaves: basal; pinnately compound,
1 1/2″ by 1 1/2″; leaf stalks 2″ to 3″
long, hairy; leaflets deeply lobed into
two or three segments; first leaves
may be undivided with heart-shaped
bases; long-hairy above and below.

Inflorescence: one or two flowers per
flower stalk; stalks hairy, sometimes
with a small leaf.

Flowers: petals yellow, 1/2″ long, curving upward; many stamens and pistils; sepals oval, 3/16″ long; flowering
from late April to mid-May.

Fruits: one-seeded, flat, circular,
1/16″ long, with a curved point at
one end; the numerous fruits of
each flower forming a ball 1/4″
in diameter; fruiting begins in
early May.

Habitat: infrequent, mostly on dry,
rocky, or sandy prairies; also on
mesic prairies and sometimes in
open woods.

variations

Bristly crowfoot
Bristly buttercup
Ranunculus pensylvanicus L. f.

Stem: perennial; 1 1/2′ to 2 1/2′ tall;
some branching above; stiff yellowish hairs.

Leaves: alternate; divided into threes,
then sharply lobed into threes again;
blades 2 1/2″ by 1 1/2″, long-hairy
above and below; leaf stalk 3″ to 4″
long, stiff-hairy with clasping bases.

Inflorescence: clusters of two or three
flowers on separate flower stalks
from the upper leaf axils, the leaves
sessile, three-divided.

Flowers: petals white, 1/4″ long;
numerous stamens and pistils;
sepals 1/4″ long, hairy; flowering
from mid-July to early August.

Fruits: one-seeded, flattened, more or
less circular, 1/8″ long, with a short
beak; numerous fruits clustered on
an elongate base (receptacle), the
clusters 3/8″ long by 3/16″ in diameter; fruiting begins in late July.

Habitat: very infrequent on low wet
prairies and in marshes.

Purple meadow-rue
Thalictrum dasycarpum Fischer &
Ave-Lall.

Stem: perennial; 3′ to 5′ tall; branching
in the inflorescence; smooth.

Leaves: alternate; pinnately compound
with the lower leaflets divided into
threes and the upper leaflets three-
lobed; blades 4″ by 3″ and smaller;
leaf stalks 1″ to 2″, hairy; leaflets
smooth above and hairy below.

Inflorescence: many tiny flowers on
much-branched flower stalks from
the stem tip and upper leaf axils;
cone-shaped, 8″ to 12″ long.

Flowers: without petals and the sepals
dropping early in flowering; on
male plants numerous, white stamens, 1/4″ long; on female plants
about ten white pistils per flower,
each pistil 3/16″ long with a swollen
ovary near the base and a flattened
stigmatic surface above; flowering
from mid- to late June.

Fruits: one-seeded; 1/8″ long by 1/16″
in diameter with the persistent style
forming a hook at the tip, dark
brown; usually remaining attached
to the plant until late summer; fruiting begins in late June.

Habitat: common in moist soils on
prairies, in open woods, on roadsides, and in other open places.

RHAMNACEAE
Buckthorn Family

Redroot
Ceanothus herbaceus Raf.
   C. ovatus Desf. (misapplied)

Stem: shrub; 2′ to 3′ tall; wide-spreading; new growth yellow-hairy.

Leaves: alternate; oval with rounded bases and pointed tips; three main veins from the base of the blade; blades 1 3/8″ by 1/2″ and larger; leaf stalks 1/4″, hairy; fine-toothed margins; very hairy above and below.

Inflorescence: dense clusters of flowers on branching flower stalks from the upper leaf axils; clusters about 4″ long.

Flowers: petals white, 1/16″, in-curved; sepals 1/32″, oval; flower parts on a broad basal disk; flowering from mid-May to early June.

Fruits: three-lobed capsules, 1/8″ across; opening explosively when mature to eject three seeds; seeds 3/32″ long, oblong with light brown, glossy surface; fruiting begins in early June; seeds shed beginning in mid-June.

Habitat: common on dry, upland, and Loess Hills prairies in western Iowa; occasionally in dry, open woods; very infrequent in eastern Iowa.

New Jersey tea
Ceanothus americanus L.

Stem: shrub; 2′ tall, wide-spreading; long-hairy on new shoots.

Leaves: alternate; oval with rounded bases and pointed tips, three main veins from the base of the blades; blades 2 1/2″ by 1 1/4″, the tips often curling upward; leaf stalks 3/8″; small teeth on the margins; long-hairy above and below, especially on the veins.

Inflorescence: tight clusters of small flowers on much-branched flower stalks from the upper leaf axils; above the leaves; clusters up to 4″ long; flower stalks very hairy.

Flowers: petals white, 1/16″, in-curved; sepals 1/32″, oval; flower parts on a broad basal disk; flowering from late June to mid-July.

Fruits: three-lobed capsules, 1/8″ across; opening explosively when mature to eject three seeds; seeds 3/32″ long, oblong, with dark brown glossy surface; fruiting begins in mid-July; seeds shed in September.

Habitat: common on dry, upland prairies and in open woods.

ROSACEAE
Rose Family

Wild strawberry
Fragaria virginiana Duchesne

Stem: perennial; very short with basal
leaves.

Leaves: basal; divided into three
leaflets; blades 3″ by 1 1/2″; leaf stalks
2″ to 6″ long, long-hairy; sharp marginal teeth on the leaflets, the end
tooth of the middle leaflet usually
shorter than those adjacent; few
hairs above and below.

Inflorescence: few-flowered cluster on
a flower stalk, which does not rise
above the leaves.

Flowers: petals white, 1/4″ long on a
1/4″ wide floral cup; sepals 1/4″ long,
alternating with bracts; many yellow
stamens; flowering from early to
late May.

Fruits: fleshy red receptacle, dome-
shaped, 3/8″ long with one-seeded
fruits embedded in the surface;
fruiting begins in mid-May.

Habitat: common on dry to moist
prairies, in open woods, on roadsides, and in other open places.

Prairie smoke

Long-plumed purple avens

Geum triflorum Pursh

Stem: perennial; very short with basal leaves.

Leaves: basal; pinnately divided; blades 6″ by 2 1/2″; leaf stalks 2″, long-hairy; leaflets above the middle of the blade longer, outer ends of the leaflets long-toothed; hairy above and below.

Inflorescence: umbel of three flowers at the end of the 6″ to 12″ hairy flower stalk.

Flowers: petals white, nearly hidden by the red calyx; calyx limbs 1/2″ long, broad-pointed, alternating with longer (5/8″) narrow bracts which elongate in fruit; stamens and pistils many; flowering from early to late May.

Fruits: many; one-seeded, 1/16″ long with the elongate hairy style attached; style reaching 1 1/2″ in length, producing a ball of hairs extending from the calyx in fruit; fruiting begins in mid-May.

Habitat: infrequent on moist prairies.

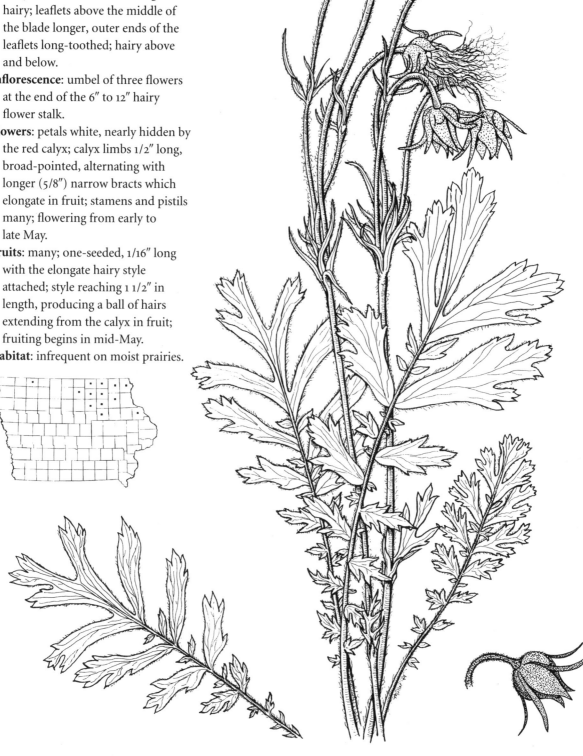

Tall cinquefoil
Potentilla arguta Pursh

Stem: perennial; 2 1/2′ to 4′; unbranched; hairy.

Leaves: alternate; pinnately compound; blades 5″ by 2 1/2″, widest above the middle; leaf stalks 2″ to 3″ long, typically five pairs of leaflets plus the end leaflet; margins of leaflets double-toothed; upper leaves sessile, with fewer leaflets; hairy above and below.

Inflorescence: clusters of flowers at ends of branching hairy flower stalk; inflorescence 3″ to 5″ long.

Flowers: petals cream, 1/2″ long; calyx 1/2″ long, lobes pointed with shorter bracts between the lobes; stamens and pistils many; flowering from early to late July.

Fruits: many, one-seeded, 1/32″ long, brown; often remaining in the dry, nearly closed calyx; fruiting begins in late July.

Habitat: frequent on dry to mesic prairies.

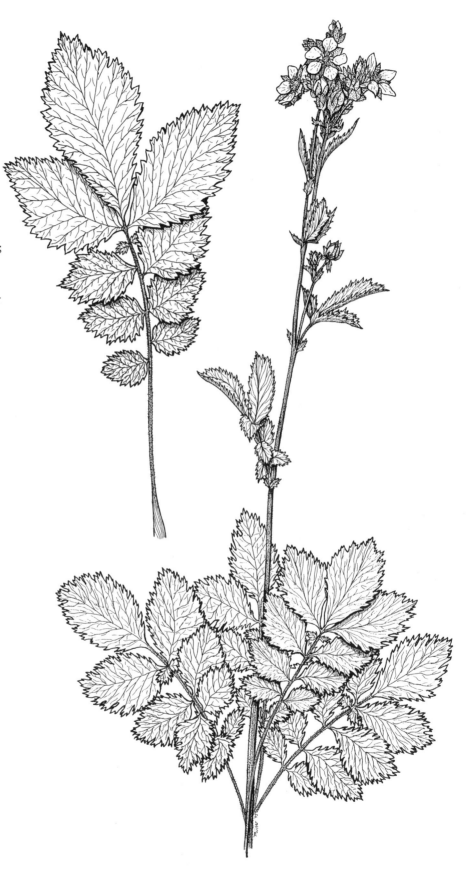

Sunshine rose
Prairie rose
Rosa arkansana Porter
R. suffulta Greene

Stem: perennial; 2′ tall shrub; prickly.
Leaves: alternate; pinnately compound; blades 5″ by 2″; leaf stalks
1/2″ to 1″ long, smooth; seven to
nine leaflets per leaf, 7/8″ by 1/2″,
toothed margins; smooth above and
below.
Inflorescence: several flowers at the
end of a flower stalk from an upper
leaf axil; corymbiform.
Flowers: petals pink to red-pink, 5/8″
long; sepals lance-shaped, 1/2″ long;
many yellow stamens; petals and
sepals attached to the top of the
round ovary, 1/4″ in diameter; flowering from early June to early July.
Fruits: a "hip" 3/8″ in diameter; red or
brown; attached sepals not curved
backward; one-seeded fruits within
the hip 5/16″ long, tan, hairy on one
side; fruiting begins in late June.
Habitat: common on upland prairies;
also on sandy prairies, on roadsides,
and in other open places.

Sunshine rose hybridizes extensively
with pasture rose, *Rosa carolina*, sometimes making determinations difficult.

Meadow rose

Smooth rose

Rosa blanda Aiton

Rosa blanda is similar to *R. arkansana* but the *dark red* stems are mostly smooth with *few prickles*. The leaves usually have *fewer leaflets* (5–7) and have *few hairs* on the leaflets. The inflorescence, flowers, and fruits are similar. Flowering is from late May to late June, and fruiting begins in mid-June. *R. blanda* is common on *low* to upland prairies and is also in open woods. It is less common in southern Iowa.

R. blanda

Pasture rose

Carolina rose

Rosa carolina L.

Rosa carolina is similar to *R. arkansana* with the stems more or less prickly. The leaves are similar except *the leaflets are fewer* (5–7), *smaller* (5/8″ by 1/2″), and *hairy above and below*. The inflorescence, flowers, and fruits are similar except the *flower stalks and sepals have glandular hairs* and the *sepals are reflexed in fruit*. Flowering is from mid-June to mid-July, and fruiting begins in early July. *R. carolina* is common on upland to dry prairies in eastern Iowa and sometimes in open woods and is very uncommon in western Iowa.

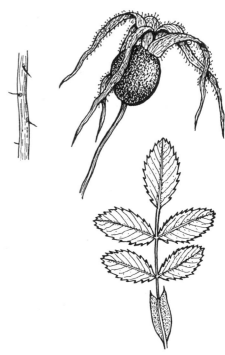

R. carolina

Meadowsweet
Spiraea alba Du Roi

Stem: perennial; 3′ shrub with numerous stems from the same root crown; smooth, tan bark.

Leaves: alternate; linear-oval with abrupt-rounded bases and pointed tips; 2″ by 1/2″; leaf stalks 1/8″ long, smooth; margins toothed; smooth above and below.

Inflorescence: conic or columnar cluster of flowers on a branching flower stalk at the stem tip; 3″ to 5″ long.

Flowers: petals white, 3/32″ long on a floral cup 1/8″ across; sepals triangular, 1/32″ long; many stamens and five pistils; flowering from early July to mid-August.

Fruits: five seedpods radiate from the center, 3/32″ long; splitting on upper side to release the tiny seeds; fruiting begins in mid-July.

Habitat: common on moist prairies; becoming very infrequent in the western and southern part of its range in Iowa.

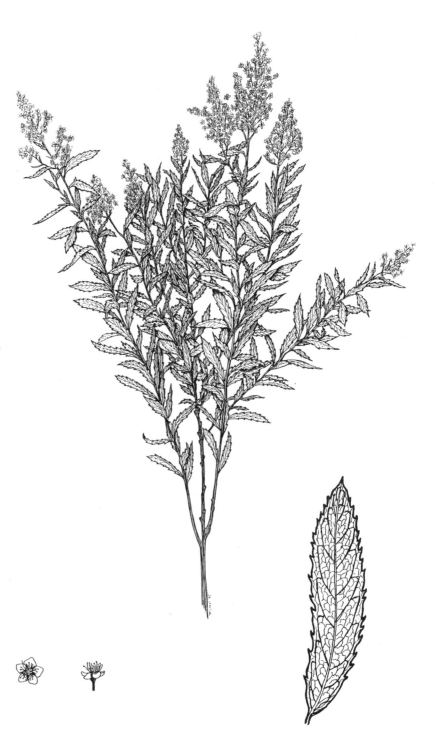

RUBIACEAE
Bedstraw Family

Northern bedstraw
Galium boreale L.

Stem: perennial; 1′ to 2′ tall; much branched; hairs on the angles of the stem.

Leaves: whorled, mostly four leaves at each node; linear-oval with pointed tips; 1″ by 1/16″; three-veined near the bases; sessile; stiff hairs on the leaf margins and midribs below, smooth above.

Inflorescence: much branched; many flowers; cylindrical to conical, 2″ to 4″ long.

Flowers: corolla white, four lobes, 1/16″ long; ovary below the corolla (inferior ovary); no calyx; flowering from mid-June to mid-July.

Fruits: two-lobed, 1/16″ long; hairy; fruiting begins in early July.

Habitat: common on moist to mesic prairies; also in open woods and on roadsides; infrequent to the west and south.

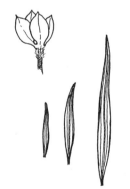

G. boreale

Wild madder
Galium obtusum Bigelow

Galium obtusum is similar to *G. boreale* except *more sprawling*. The stems are up to 2′ long, ridged, but *without hairs*. The leaves are *shorter* and *with only one vein*. The leaves are congested below inflorescence at flowering. There are *fewer flowers* in the little-branched inflorescence. The corolla is *smaller* (1/32″ long). The fruits are *smooth*, 3/32″ in diameter, and often with only one seed maturing in each fruit. Flowering is from early June to early July, and fruiting begins in mid-June.

G. obtusum is common on low, *moist prairies and in marshes*, is sometimes in open woods, and is less frequent to the west and south.

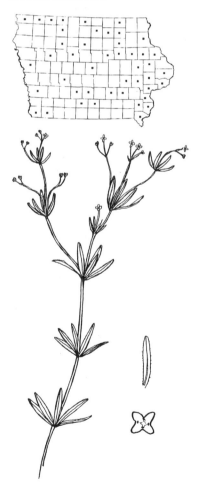

G. obtusum

SALICACEAE
Willow Family

Prairie willow
   Upland-willow
   Salix humilis Marsh.

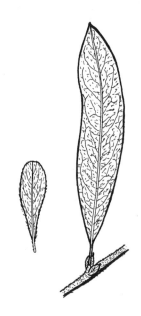

Stem: perennial shrub; to 5′ tall; the
   new bark hairy.

Leaves: alternate; oval-linear with
   tapering bases and pointed tips;
   blades 2″ to 4″ by 1/2″; petioles 3/16″,
   hairy; early season leaves smaller
   and wider above the middle; basal
   leaflets (stipules) oval-pointed,
   3/16″ long, sometimes not present;
   margins smooth or sometimes
   undulating; green-hairy above,
   gray-hairy below.

Inflorescence: plants male or female;
   fuzzy catkins, 1″ by 1/4″ in diameter,
   near the ends of the branches;
   appearing before the leaves.

Flowers: flowers without calyx or
   corolla; tiny bracts and numerous
   hairs attached at the base of each
   flower giving the inflorescence a
   fuzzy appearance; one pistil in the
   female flowers; several stamens in
   each male flower; flowering from
   early to late April.

Fruits: hairy capsules, 5/32″ long; splitting into two halves, releasing tiny
   seeds covered with long cottony
   hairs; fruiting begins in late April.

Habitat: common on dry to moist
   prairies, sometimes in open woods.

Pussy willow
Salix discolor Muhl.

Salix discolor is similar to *S. humilis*, but the *stems are more hairy*. The leaves are *somewhat larger*, 2 3/4″ by 1 1/2″, with the early season leaves 1 1/2″ by 5/8″. The leaf margins are *mostly toothed*, and the basal leaflets (stipules) have *heart-shaped bases*, about 1/4″ long (sometimes not present). The leaves are *white-hairy below* and green-hairy above. The inflorescence, flowers, and fruits are similar. Flowering is from early to late April, and fruiting begins in late April. *S. discolor* is common on *moist prairies* and in *open, wet areas*, becoming less common to the west and south.

Shrub willow
Salix petiolaris Smith

Salix petiolaris is similar to *S. humilis* except it is *usually shorter* (to 4′) and the stems are *reddish and smooth*. The leaf margins are *toothed*. The leaves are *larger* (to 3″ by 3/4″), and have *sharper points*, with *no basal leaflets (stipules)*. The leaves appear *with the flowers*. The inflorescence, flowers, and fruits are similar, but the fruits are *longer* (to 1/4″). Flowering is from early to late April, and fruiting begins in mid-May. *S. petiolaris* is common in northeast Iowa on *moist prairies and in marshes*.

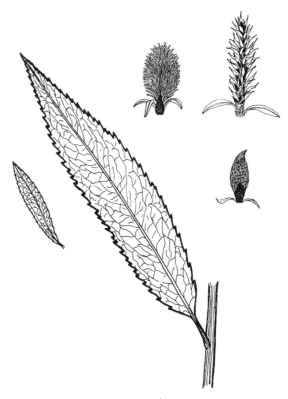

S. petiolaris

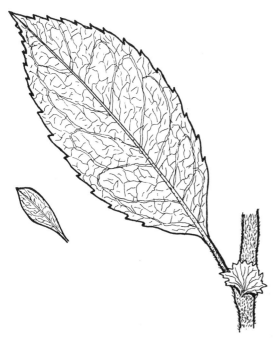

S. discolor

S. petiolaris

SANTALACEAE
Sandalwood Family

Bastard toadflax
Comandra umbellata (L.) Nutt.

Stem: perennial; 6″ to 12″ tall; sometimes branched; smooth; parasitic
on roots of other plants.

Leaves: alternate; oval with pointed
tips; 1″ by 1/4″; sessile; crowded;
reticulate veins; smooth above and
below; nonflowering plants produce
progressively smaller leaves toward
the stem tip.

Inflorescence: few flowers at the tips
of branched flower stalks from the
upper leaf axils and stem tip;
corymbiform.

Flowers: five white sepals, 1/16″ long
on a floral cup 1/16″ wide; flowering
from mid-May to early June.

Fruits: light brown, one-seeded
berries, 3/16″ in diameter; fruiting
begins in early June.

Habitat: common on upland, dry
prairies; sometimes on lowland
prairies and in open woods.

SAXIFRAGACEAE
Saxifrage Family

Alumroot
Heuchera richardsonii R. Br.

Stem: perennial; with basal leaves and leafless flower stalk.

Leaves: basal; round, lobed with palmate veins; blades 2 1/2″ by 3″; leaf stalks 3″ to 6″, with stiff hairs; leaf margins shallow-lobed with each lobe toothed; few, stiff hairs below, mostly on the veins, fewer hairs above.

Inflorescence: branched; drooping flowers on a long flower stalk, 1 1/2′ to 2 1/2′ tall, hairy.

Flowers: white to greenish, with an elongate, oblique-margined floral tube, 3/8″ long; sepals and petals 1/16″ long, alternating on the margin of the floral tube; flowering from late May to mid-June.

Fruits: capsules, 5/16″ long, wide at the base and tapering to the tip; splitting at the tips into two recurved points; seeds tiny, remaining in the upright capsule; fruiting begins in mid-June.

Habitat: common on upland to dry prairies; sometimes on moist prairies and in open woods.

Swamp saxifrage

Saxifraga pensylvanica L.

Stem: perennial; with basal leaves and tall flower stalk.

Leaves: basal; oval, tapering at both ends; blades 8″ by 2 1/2″; petioles 1 1/2″, flat, hairy, sheathing the stem at the base; few long hairs above, short below.

Inflorescence: dense masses of small flowers at the tips of much-branched flower stalks, 1 1/2′ tall; smaller flower masses below on branches from the axils of bracts 1/2″ by 1/16″.

Flowers: petals white, narrow, 1/16″ long, alternating with green sepals, 1/32″ long on a broad floral tube, 1/8″ across; flowering from mid-May to early June.

Fruits: two divergent seedpods, 1/8″ long, sepals reflexed; fruiting begins in late May; during fruiting the flower stalk elongates to 3′ and the flowers become less massed.

Habitat: frequent on wet prairies; also in moist, open woods; very uncommon at the west and south edges of its range in Iowa.

SCROPHULARIACEAE
Figwort Family

Indian paintbrush
Castilleja coccinea (L.) Sprengel

Stem: perennial; 1′ to 2′ tall; unbranched; with stiff spreading hairs.

Leaves: alternate; linear, 2″ by 1″ with three narrow, linear, spreading lobes at the tips; sessile; long-hairy above and below.

Inflorescence: raceme; crowded flowers, each with a three-lobed, red (sometimes yellow) bract below each flower.

Flowers: calyx red, 5/8″ long, tubular, two-lipped, hairy; the corolla two-lipped, extending beyond the calyx; flowering from mid-May to early June.

Fruits: capsules, two-lipped, 5/16″ long; developing within the calyx; the colored bracts dropping in fruit; fruiting begins in late May.

Habitat: infrequent on moist prairies; often occurring at the woodland-prairie border.

Castilleja coccinea is a root parasite on other plants. The yellow color phase is often found on prairies in north-central Iowa.

Downy painted cup
Castilleja sessiliflora Pursh

Stem: perennial; 1′ to 1 1/2′ tall; little
branching; sometimes several stems
from one root crown; densely hairy.

Leaves: mostly alternate, crowded; linear, 1″ by 1/8″; usually with three
diverging lobes at the tips; sessile;
hairy above and below.

Inflorescence: raceme; with a three-
lobed bract below each flower,
nearly equaling the flower in length;
flowers crowded; inflorescence up
to 8″ long.

Flowers: calyx pale yellow, tubular,
hairy, 1″ long, two-lipped, upper lip
1/4″ longer than the lower; corolla
two-lipped, protruding 5/8″ beyond
the calyx; flowering from mid-May
to mid-June.

Fruits: capsules two-lipped, 1/2″ to
5/8″ long by 1/4″ in diameter; developing within the calyx.

Habitat: common on dry to mesic
prairies, especially on the Loess Hills
prairies of western Iowa; very infrequent in eastern Iowa.

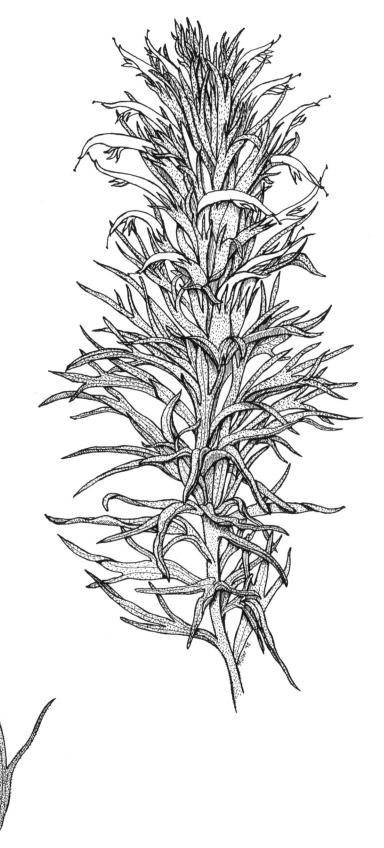

Lousewort
Pedicularis canadensis L.

Stem: perennial; 1′ to 1 1/2′ tall; sometimes several stems from one root crown; hairy.

Leaves: usually alternate; pinnately lobed, each lobe again lobed; lower leaves 3″ by 1″; leaf stalks very hairy; smooth above, few hairs below.

Inflorescence: raceme; small-hairy, lobed leaves below each flower; flowers crowded.

Flowers: corolla yellow, tubular, 1″ long, upper lip hooked downward, lower lip shorter, divided into three lobes; calyx 5/16″ long, two-lipped; flowering from early to late May.

Fruits: capsule, 5/16″ long, opening into four segments; lower part of calyx remaining intact, enclosing base of fruit; inflorescence elongating in fruit, separating the fruits; fruiting begins in mid-May.

Habitat: frequent on dry, sandy, and rocky prairies; also in open woods on thin soils.

Swamp lousewort
Pedicularis lanceolata Michx.

Stem: perennial; 2′ to 3′ tall; some
branching above the middle;
sparsely hairy below.

Leaves: mostly opposite; oval-linear, 3″
by 3/4″; sessile; margins lobed 1/8″
deep, the lobes finely toothed; few,
long hairs above and below.

Inflorescence: raceme; flowers
crowded; leafy bracts below each
flower 3/8″ long, finely toothed and
with stiff, marginal hairs.

Flowers: corolla yellow, tubular, 7/8″
long, upper lip curved downward;
calyx two-lipped, 1/4″ long, hairy on
the inside at the base of each lip;
flowering from mid-August to mid-
September.

Fruits: capsules, 3/8″ long, two-lipped;
maturing within the calyx; fruiting
begins in late August.

Habitat: frequent on wet to moist
prairies, in marshes, and in wet,
open woods; becoming very infrequent in the western and southern
parts of its range in Iowa.

Large-flowered beardtongue
Penstemon grandiflorus Nutt.

Stem: perennial; 2′ to 3′ tall;
  unbranched; smooth with waxy,
  bluish cast on stem and leaves.
Leaves: opposite; lower round-oval,
  2″ by 1 3/8″, upper slightly chordate,
  1″ by 1″; sessile to somewhat clasping; smooth above and below.
Inflorescence: raceme with two flowers per node; opposite bracts at each
  node, 5/8″ long, oval.
Flowers: corolla purple to lavender,
  two-lipped, narrow, funnel-shaped,
  1 3/8″ long; calyx five-lobed, 5/16″
  long; flowering from late May to
  mid-June.
Fruits: capsules; 7/8″ long, widest near
  the base, opening at the tip by four
  segments; seeds squarish, 3/32″ long;
  fruiting begins in early June.
Habitat: common on Loess Hills prairies in western Iowa; very infrequent
  elsewhere on dry, sandy prairies and
  stabilized sand dunes.

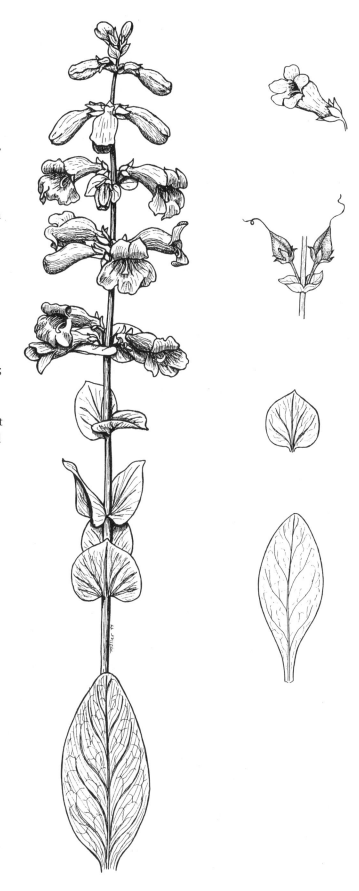

Pale beardtongue
Penstemon pallidus Small

Penstemon pallidus is similar to
P. grandiflorus except *shorter* (1′–3′)
and with *hairy stems*. The *leaves are
narrower* (2″ by 3/4″), taper to pointed
tips, and are *long-hairy* above and
below. The inflorescence has *flowers on
axillary branches* from the upper
nodes. The *flowers have white corollas*
and are *shorter* (5/8″). The *fruits are
smaller* (5/16″) and *split into two halves.*
Flowering is from mid-May to mid-
June, and fruiting begins in early June.
P. pallidus is frequent in *dry, sandy soils*
on prairies, in open woods, and in
woodland openings in southeastern
Iowa but elsewhere is very infrequent.

P. digitalis

Foxglove penstemon
Foxglove beardtongue
Penstemon digitalis Nutt.

Penstemon digitalis is about the same
height as *P. grandiflorus*, but the leaves
are *somewhat larger and narrower*
(5″ by 3/4″). The inflorescence has *flowers on axillary branches* from the upper
nodes. The flowers are similar, but the
corolla is shorter (7/8″) with *hairs on
the outside.* The *calyx is shorter* (3/16″).
The fruits are similar but *smaller*
(1/4″ long). Flowering is from early to
late June, and fruiting begins in mid-
June. *P. digitalis* is frequent in *moist,
sandy soils* on prairies and in other
open places in southeastern Iowa,
becoming very infrequent to the north
and west.

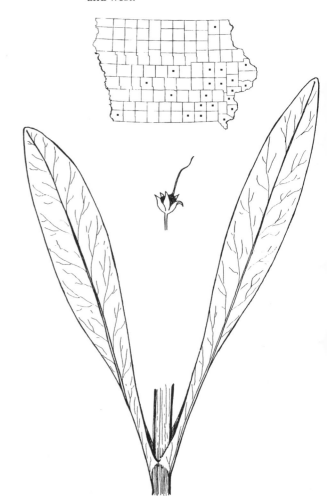

P. digitalis

P. pallidus

Figgwort
Scrophularia lanceolata Pursh

Stem: perennial; 3′ to 5′ tall; squarish; unbranched; smooth.

Leaves: opposite; oval with abrupt bases and pointed tips; blades 3″ by 1 3/8″; "winged" leaf stalks 1″, smooth; margins double-toothed; smooth above and below.

Inflorescence: panicles from the upper stem nodes, 6″ to 8″ long, cylindrical; flowers widely spaced; hairy (glandular).

Flowers: corolla green (to reddish brown), cup-shaped, 5/8″ long, with two lobes above and three below (1/16″ to 1/8″ long); sterile stamen at the back (within the flower), flat, broad with a blunt, yellow head; calyx small, 1/8″ long; flowering from late May to late June.

Fruits: capsules, 1/4″ long, widest at the base, tapering to a point; splitting into two halves; fruiting begins in mid-June.

Habitat: frequent on dry and sandy prairies, also on mesic prairies, in open woods, on pastures, and on roadsides in northeastern Iowa; becoming less frequent to the south and west.

Another figwort, *Scrophularia marilandica* L., with a dark purple or brown sterile stamen, is usually restricted to open woods.

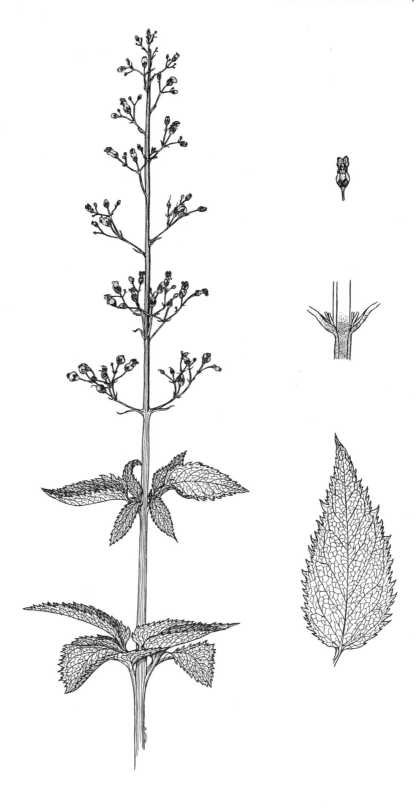

Culver's root
Veronicastrum virginicum (L.) Farw.

Stem: perennial, 3′ to 4′ tall; unbranched; hairy.

Leaves: whorled, five, occasionally four, at each node; oval, tapered to both ends; 3″ by 5/8″; leaf stalks very short, 1/8″; toothed margins; smooth above and below.

Inflorescence: raceme, several at the ends of flower stalks from the stem tip and upper leaf axils; flowers very densely packed; flowering begins at the base.

Flowers: corolla white, tubular, 3/16″ long with small lobes; stamens projecting well beyond the corolla, yellow anthers; calyx light green, deeply lobed, 1/16″ long; flowering from early July to early August.

Fruits: capsules, 1/8″ long, splitting into two halves; seeds very tiny; fruiting begins in mid-July; plants turn black as the fruits mature in August.

Habitat: common on moist prairies and in open woods, into somewhat disturbed sites.

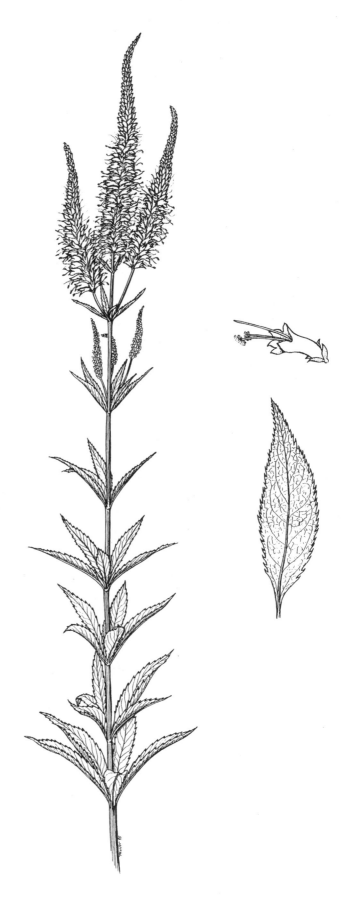

SOLANACEAE
Nightshade Family

Virginia ground cherry
Physalis virginiana P. Miller

Stem: perennial; 1′ to 2′ tall; branching; lightly hairy.

Leaves: alternate; oval, tapering to both ends; blades 1 1/2″ by 1/2″; leaf stalks 3/4″ long, hairy; margins smooth to undulate to irregularly toothed; hairy above and below.

Inflorescence: single flowers at the upper leaf axils, drooping.

Flowers: corolla yellow, funnel-shaped, 3/4″ across; calyx 3/8″ long with pointed lobes; flowering from early June to early July.

Fruits: orange to red berries, 1/4″ in diameter, enclosed within the inflated calyx, 3/4″ long; fruiting begins in mid-June.

Habitat: frequent on dry to moist prairies and in open woods; often associated with disturbed habitats.

Ground cherry
Clammy ground-cherry
Physalis heterophylla Nees

Physalis heterophylla is similar to *P. virginiana* except the stem is *sticky-hairy*. The *leaves are wider* (1 1/2″ by 1 1/4″), and the bases are more rounded. The leaf margins have *large, shallow teeth*, and the leaves are *more hairy on the upper surface*. The inflorescence is similar, but the flowers are *somewhat smaller* (5/8″ across). The fruits are similar but are *green to yellow* in color. Flowering is from mid-June to *mid-August*, and fruiting begins in early July. *P. heterophylla* is common on mesic and sandy prairies, on roadsides, in open woods, and in other open, disturbed places.

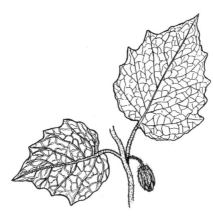

P. heterophylla

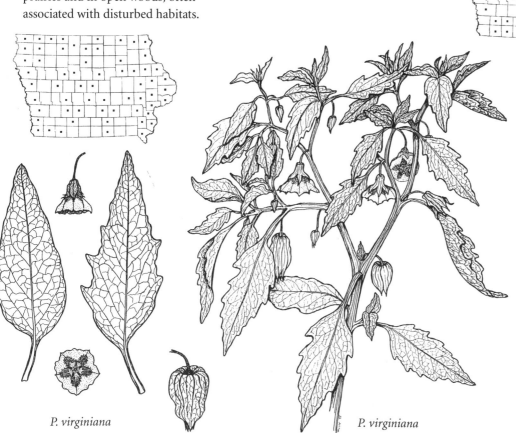

P. virginiana

P. virginiana

URTICACEAE
Nettle Family

Bog clearweed
Pilea fontana (Lunell) Rydb.

Stem: annual; 1′ to 1 1/2′ tall; usually
not branching; vascular bundles visible in transparent stem; smooth or
a few long hairs.

Leaves: opposite; oval with blunt-
tapered bases and pointed tips; 1″ by
3/4″; leaf stalks 1/2″, with a few hairs;
margins round-toothed; smooth
above and below.

Inflorescence: short branched
flower stalks from the stem tip
and upper leaf axils; flowers very
inconspicuous.

Flowers: sepals green, 1/32″ long; stamens longer than the sepals; no petals; flowers male or female; flowering from mid- to late July.

Fruits: one-seeded, flat, purple-black;
1/16″ long; fruiting begins in early
August.

Habitat: very uncommon in fens and
seepage areas to moist prairies.

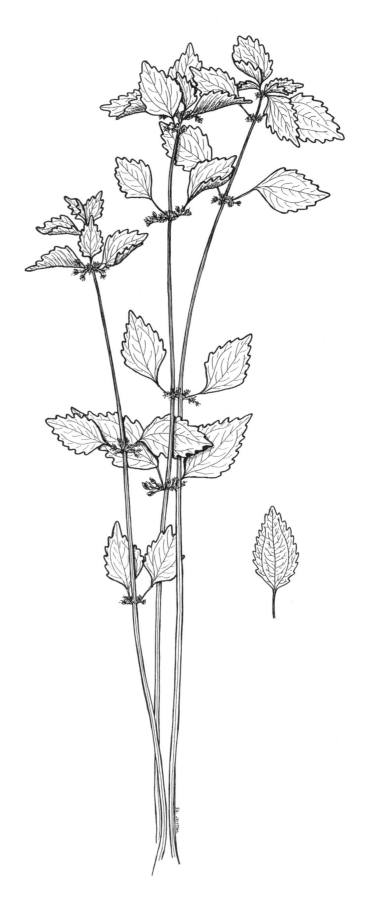

VERBENACEAE
Vervain Family

Blue vervain
Verbena hastata L.

Stem: perennial; 3′ to 5′ tall; unbranched to the inflorescence; stem square; lightly hairy; often very short branches, with 1/2″ to 1″ leaves developing in axils of stem leaves.

Leaves: opposite; oval, with rounded bases, tapering to sharp tips, sometimes with paired lobes near base of blades; blades 3″ long by 1″ wide; leaf stalks 1/2″ long, hairy; margins coarsely toothed; hairy above and below.

Inflorescence: numerous racemes on flower stalks at the stem tip and from the upper leaf axils; flowers crowded; raceme elongating during development, from 1″ as flowering begins to 4″ as fruits mature.

Flowers: corolla blue (sometimes white), 1/4″ long, dropping soon after opening; calyx tubular with short, pointed lobes; flowering from early July to mid-August.

Fruits: four nutlets develop within the calyx; 3/32″ long; fruiting begins in mid-July.

Habitat: common on moist prairies, in open woods, on stream banks, and in other moist, open places; sometimes on drier, open sites; tolerates disturbance.

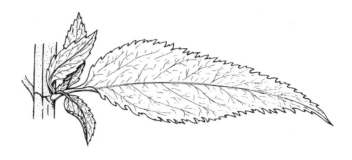

Hoary vervain
Verbena stricta Vent.

Stem: perennial; 1 1/2′ to 2 1/2′ tall; little branched; stem square; hairy.

Leaves: opposite; oval with rounded tips and tapering bases; 1 1/2″, occasionally up to 4″, by 3/4″; sessile; margins with small teeth; very hairy above and below.

Inflorescence: a raceme, one or a few from the stem tip and upper leaf axils; flowers crowded; raceme elongating greatly during development, to 6″ as fruits mature.

Flowers: corolla deep blue, tubular, two-lipped, 1/8″ beyond the calyx, dropping soon after opening; calyx tubular, 3/16″ with small teeth; flowering from mid-June to early September.

Fruits: four nutlets developing within the calyx; fruiting begins in late June.

Habitat: common on mesic to dry prairies, pastures, and roadsides to open woods; most abundant in disturbed places.

V. stricta

V. stricta

Narrow-leaved vervain
Verbena simplex Lehm.

Verbena simplex is *shorter* than *V. stricta* (1′ to 1 1/2′ tall) and is *not as hairy on the stem and leaves*. There is a tendency to produce very short branches in the axils of the stem leaves. The leaves are *narrow* (1 1/2″ by 1/4″) and toothed on the upper half, with *few hairs above and below*. The flowers are not as crowded in the inflorescence, and the lobes of the corolla are *fringed*. Flowering is from mid-June to late August, with fruiting beginning in late June. *V. simplex* is frequent on *sandy prairies and open bottomland* and is able to tolerate disturbance.

V. simplex

White vervain
Verbena urticifolia L.

Stem: perennial; 3′ to 4′ tall; unbranched; stem square; hairy.

Leaves: opposite; oval, with rounded bases and tapering, sharp tips; blades 4″ by 1 3/4″; leaf stalks 1/2″, few, stiff hairs; margins toothed; hairy.

Inflorescence: racemes on flower stalks from the stem tip and upper leaf axils; flowers crowded at the tip of the raceme but elongating in fruit; up to 6″ long at the end of the season.

Flowers: corolla white, exceeding the calyx by 1/16″, dropping soon after opening; the calyx tubular, 3/32″ long, with sharp lobes, hairy; flowering from mid-July to early August.

Fruits: four nutlets, just exceeding the calyx; fruiting begins in late July.

Habitat: common in open woodland on upland to lowland sites; also on low prairies, pastures, and roadsides; tolerates some disturbance.

VIOLACEAE
Violet Family

Prairie violet
Viola pedatifida G. Don

Stem: perennial; very short with basal
 leaves.

Leaves: basal; deeply lobed into linear
 segments, each segment again lobed;
 blades 1 1/4″ by 1 1/2″ and larger; leaf
 stalks 1″ to 4″ long, smooth; hairs on
 the margins and veins.

Inflorescence: flowers solitary on
 smooth flower stalks from the base
 of plant; open flowers on flower
 stalks slightly above the leaves;
 nonopening flowers on shorter
 flower stalks.

Flowers: petals violet, about 5/8″ to
 3/4″ long, with hairs near the base
 of the lower three petals; sepals
 1/4″ long; base of flower curved
 backward under flower stalk forming a spur containing nectar-
 producing glands; flowering from
 early to late May.

Fruits: capsule 7/16″ long, opening into
 three segments; open flowers seldom
 producing seeds; nonopening flowers on short stalks below the leaves
 setting fruit through self-pollination;
 fruiting begins in late May.

Habitat: frequent on mesic to dry
 prairies; also on moist prairies.

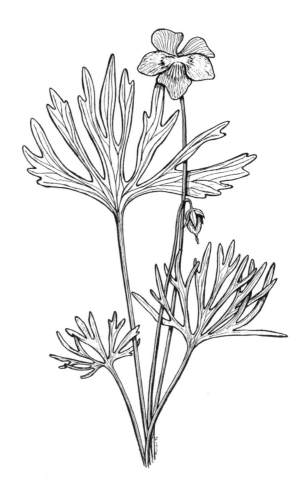

Hybrids between *Viola* species are not
uncommon, and identification of vio-
lets is sometimes difficult.

Bird's-foot violet
Viola pedata L.

Viola pedata is quite similar to *V. peda-tifida* except the leaves are *lobed into narrower segments with fewer secondary lobes, especially on the center lobe.* The first leaves of the season are less divided and smaller (3/4″ long). The petals are *lighter in color, the stamens and style protrude from the center of the flower,* and *the petals are without hairs.* Flowering is from early to late May, and fruiting begins in late May. *V. pedata* is frequent on dry, rocky, and sandy prairies and is very uncommon in the western part of its range in Iowa.

V. pedata

V. pedata

Common blue violet
Viola pratincola Greene
V. papilionacea Pursh

Viola pratincola is similar to *V. peda-tifida* but has "typical" *heart-shaped leaves* (sometimes with rounded tips) with blades up to 2″ by 3″ and *smooth above and below.* The flowers are raised to about the same level as the leaves. The flowers are similar (petals 3/4″ long), *with hairs on the lower petals.* The fruits are similar. Flowering is from late April to late May, and fruiting begins in late May. *V. pratincola* is frequent on moist prairies and in open woods as well as in lawns.

V. pratincola

ANGIOSPERMS: MONOCOTYLEDONS

AGAVACEAE
Yucca Family

Soapweed
Yucca
Yucca glauca Nutt. ex Fraser

Stem: perennial; basal leaves and a tall flower stalk to 5′ tall.

Leaves: basal; numerous; linear, gradually tapering to a spine-tip; 1′ to 2′ by 1/2″; threads on the margins; smooth above and below.

Inflorescence: raceme at the tip of the flower stalk; raceme 6″ to 2′ long; each flower attached to the flower stalk above a bract, 1″ long, lance-shaped.

Flowers: petals and sepals white, large (1 1/2″ long), forming a cup-shaped flower; on a curved stalk, the flower facing downward; flowering from early to mid-June.

Fruits: capsule, 1 1/2″ to 2″ long by 1″ in diameter, held upright; three-chambered with many flat, black seeds; fruiting begins in late June.

Habitat: frequent on west- and south-facing Loess Hills prairies in the western tier of counties in Iowa.

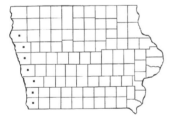

Soapweed is pollinated by night-flying moths who use the ovary of the flower as the site to lay their eggs. The young larvae eat the developing seeds.

COMMELINACEAE
Spiderwort Family

Spiderwort
Tradescantia bracteata Small

Stem: perennial; 1′ to 2′ tall; unbranched; smooth.

Leaves: alternate; linear, with sheathing base, tapering to a sharp point at the tip; 6″ to 8″ long by 1/2″ to 1″ wide; scattered hairs above, smooth below.

Inflorescence: umbel of few to many flowers on flower stalks from the stem tip and upper leaf axils; the flower stalks of the individual flowers long-hairy; the flower buds bent downward but bending upright at blooming; each flower cluster arising from two basal bracts, the bracts shorter than the stem leaves.

Flowers: petals blue to rose-purple, 1/2″ long; sepals 7/16″ long, long-hairy; the filaments of the stamens with long, blue hairs; flowering from mid-May to mid-June.

Fruits: capsule, 3/16″ long, splitting into three segments; developing within the calyx; fruiting begins in late May.

Habitat: common on dry to moist prairies; also in open woods, on roadsides, and in other open places; infrequent in southeastern Iowa.

Ohio spiderwort
Tradescantia ohiensis Raf.

Tradescantia ohiensis is similar to spiderwort except the stems are often *taller* (to 2 1/2′). The leaves are *up to 2′ long but narrower* (to 1/2″ wide) with *cottony hairs on the margins above the sheath*. The stalks of the individual flowers are *smooth*, and the sepals are *smooth* except for marginal hairs and sometimes a tuft of hairs at the tip. Fruits are similar. Flowering is from *late May to late June*. Fruiting begins in early June. *T. ohiensis* is common on *sandy prairies and in other open, sandy places*, often with some disturbance, and also in open woods.

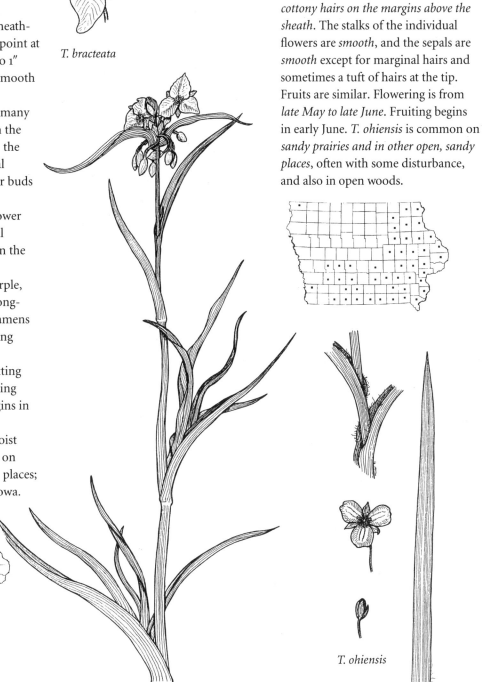

T. bracteata

T. ohiensis

T. bracteata

CYPERACEAE
Sedge Family

Sedge
Carex spp.

The sedge family is similar to the grass family with its reduced flowers and linear leaves. The family includes bulrushes, spike rushes, and nut grasses in addition to sedges. Only sedges (*Carex*) are included in this book. Sedges have one bract below each floret. The female florets have a unique structure, a perigynium, which is a membranous cylinder enclosing the pistil with its two or three styles projecting out. Male flowers have three stamens. Florets are arranged into spikes with male and female flowers segregated in one of three ways: separate spikes with the male spike taller than the female spikes, male and female flowers in the same spike with the male flowers at the base and the female flowers above, or the reverse with female flowers below male flowers in the same spike. Vegetatively sedges have leaves divided into sheath and blade as in the grasses. However, the sheath grows closed into a cylinder surrounding the stem, and leaves grow out from the stem in three rows up the stem rather than two rows as in the grasses.

The following five sedges represent more than fifty sedge species that might be found on Iowa's prairies according to Eilers and Roosa's *The Vascular Plants of Iowa*. Positive identification of the sedges is based upon many technical characteristics and requires the use of a good hand lens or a low-power microscope. If you are interested in studying the group further, consult one of the manuals listed in the bibliography.

The following key can be used to identify the five species treated here:

If the spikes are elongate to 1″ long
    Carex lasiocarpa
If the spikes are roundish
  and the florets diverge, producing a bristly ball
    and the leaf sheaths are loose and mottled or green-striped
      Carex gravida
    and the leaf sheaths are tight and not mottled or green-striped
      Carex muhlenbergii
  and the florets are appressed together
    and the leaf blades are 1/8″ wide
      Carex brevior
    and the leaf blades are 1/16″ wide
      Carex bicknellii

Perigyniums

C. lasiocarpa

C. gravida

C. muhlenbergii

C. brevior

C. bicknellii

Slender sedge
Carex lasiocarpa Ehrh.

Carex lasiocarpa is a tufted perennial. The stem is 1 1/2′ to 2 1/2′ tall, smooth, and round. The leaf blades are 12″ by 3/32″ and smooth. Male and female flowers are in separate spikes, with the male spikes above the female spikes. One or two male spikes are 1″ long by 3/32″ in diameter. One or two female spikes are 3/4″ to 1″ long by 3/16″ in diameter. A threadlike bract to 4″ long is attached at the base of the upper female spike. The perigynium is oval, tapering to two teeth at the tip, is hairy, and has three or five distinct veins. Three stigmas protrude from the perigynium. The seedlike fruit is triangular and 1/16″ long. Flowering is from mid-May to early June. Fruiting begins in late May. Fruits begin dropping in mid-June. *C. lasiocarpa* is frequent on wet prairies and marshes and in wet, sandy, open places.

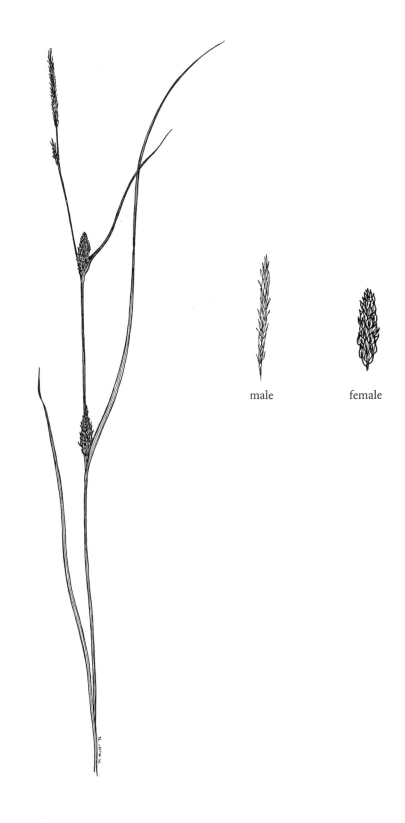

male   female

Sedge
Carex gravida Bailey

Carex gravida is a tufted perennial with a triangular stem 1 1/2′ to 3′ tall. The basal leaf sheaths are loose around the stem and mottled or green-striped. The leaf blades are 12″ by 1/4″ and smooth with the tip sharp-pointed. Usually five roundish spikes are clustered at the tip of the stem. The spikes are bristly with the florets divergent from one another. The male flowers are located at the tip of each spike. The perigynium is oval and flat with the edges winged and tapering to two 1/16″ long teeth. Two stigmas protrude from the perigynium. The seedlike fruits are flat and 1/32″ in diameter. Flowering is from mid-May to early June. Fruiting begins in early June, and fruits begin dropping in late June. *C. gravida* is frequent on moist and mesic prairies to open woods and forests.

Sedge
Carex muhlenbergii Schkuhr ex Willd.

Carex muhlenbergii is a tufted to rhizomatous perennial with a triangular stem 1′ to 2′ tall. The basal leaf sheaths are tight around the stem and green. The leaves are 10″ to 18″ by 1/8″ and gradually taper to a sharp tip. The inflorescence is lobed and 3/8″ to 1 1/4″ long, made up of several clustered spikes. The male flowers are located at the tip of each spike. Each spike has diverging florets with perigynia 3/16″ by 5/32″, oval, flat, and tapering rather abruptly to two teeth. Two stigmas protrude from the perigynium. The seedlike fruits are flat and round, are 3/32″ in diameter, and have a short beak. Flowering begins in mid-May, and fruits form in late May and June. Fruits begin dropping in late June. *C. muhlenbergii* is frequent in sandy, moist soil in open places and woodland edges.

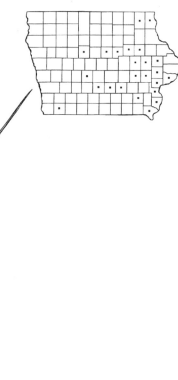

C. gravida   C. muhlenbergii

Sedge

Carex brevior (Dewey) Mack. ex Lunell

Carex brevior is a tufted perennial with a smooth stem 1 1/2′ to 2 1/2′ tall. Leaf blades are 12″ by 1/8″ and smooth. Three to five spikes are clustered at the tip of the stem. Each spike is 5/16″ long with male florets at the base of each spike. The bracts below each spike are threadlike and longer than the spike. The perigynium is oval, 1/8″ long, and flat with winged edges. The perigynium tapers to two points with two stigmas protruding. The seedlike fruits are flat, are 1/16″ in diameter, and have a prominent beak at the style end. Flowering is from mid-May to early June, and fruiting begins in late May. Fruits begin dropping in mid-June. *C. brevior* is frequent on dry to wet prairies and in open woods.

Sedge

Carex bicknellii Britton

Carex bicknellii is a tufted perennial with smooth stems 2′ to 3′ tall. The smooth leaf blades are 12″ by 1/16″. Several roundish spikes, 5/16″ long, are clustered at the tip of the stem. In each spike the male florets are at the base of the spike. The perigynium is oval, 1/8″ long, and flat with winged edges. The perigynium tapers abruptly to a two-pointed beak with two protruding stigmas. Seedlike fruits are flat and 3/32″ in diameter. Flowering is from mid-May to mid-June. Fruiting begins in late May. Fruits are held tightly through the summer. *C. bicknellii* is frequent on moist to mesic prairies and also is found on dry and sandy prairies.

C. brevior   *C. bicknellii*

IRIDACEAE
Iris Family

Blue flag
Iris shrevei L. Small
   I. virginica var. *shrevei* (Small)
   E. Anderson

Stem: perennial; 1 1/2′ to 2 1/2′ tall;
   unbranched; smooth.
Leaves: mostly basal; lance-shaped,
   tapering to a pointed tip; 1′ to 1 1/2′
   by 1″; smooth above and below.
Inflorescence: raceme of a few flowers;
   each flower above a sheathing bract,
   to 2″ long.
Flowers: petals and sepals blue,
   attached at the top of the ovary (inferior ovary); three style branches, flat
   and spreading, obscure the three
   stamens lying above the sepals;
   flowering from early to late June.
Fruits: capsule; 1 1/2″ to 3″ long by 1/2″
   in diameter; with three chambers;
   seeds flat, 1/4″ across, with corky,
   brown seed coats; fruiting begins in
   mid-June.
Habitat: frequent on marshy ground,
   pond edges, and roadsides.

As the fruits begin to develop the
flower parts curl inward and self-
digest. Insect larvae often burrow
through the ripening capsule, eating
the embryos of the seeds. Usually a few
seeds are spared.

Blue-eyed grass

Sisyrinchium campestre Bickn.

Stem: perennial; tufted, with basal leaves; flower stalk 6″ to 10″ tall; smooth.

Leaves: basal; linear, gradually tapering to a sharp tip; 3″ to 5″ by 1/16″; smooth.

Inflorescence: umbel of several flowers on a leafless flower stalk; two upright bracts of unequal length below each umbel.

Flowers: petals and sepals light violet to white, 3/8″ long, abruptly sharp-tipped; with a yellow center; flowering from early to late May.

Fruits: capsule; round, 5/32″ in diameter; three-chambered; light brown; seeds black, 1/32″ in diameter; fruiting begins in mid-May; seeds begin dropping in early June.

Habitat: common on mesic to dry prairies; also on sandy and moist prairies and in woodland openings.

LILIACEAE
Lily Family

Wild onion
Wild garlic
Allium canadense L.

Stem: perennial from a bulb; flower stalk 1′ to 1 1/2′ tall.

Leaves: mostly basal; linear, with a clasping base, tapering to a point at the tip; 6″ to 8″ by 1/16″; smooth.

Inflorescence: umbel of numerous flowers on short stalks from the tip of a 1′ to 1 1/2′ flower stalk; two or three bracts, 1/2″ long, below the umbel.

Flowers: sepals and petals pink (to white), 1/4″ long; flowers sometimes replaced by tiny bulbs, 3/8″ long, or a mixture of bulbs and flowers; flowering from early to late June.

Fruits: capsule, 1/8″ in diameter; with three chambers; fruiting begins in late June.

Habitat: frequent on mesic to dry prairies; also in open woods and in more moist habitats.

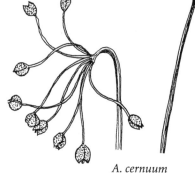

A. cernuum

A. canadense

Nodding wild onion
Allium cernuum Roth

Allium cernuum is found only in the *northeastern* part of the state where it is very infrequent on both prairies and in open woods. It has a *bend or crook* in the flower stalk near the tip and produces *only flowers* in the inflorescence. Flowers are pink or rose (sometimes white), and blooming occurs in *mid-July to early August*. It occurs on upland prairies and in open woods.

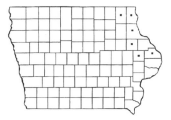

Wild prairie onion
Allium stellatum Nutt. ex Ker-Gawl.

Allium stellatum occurs only in *northwest* Iowa. It has leaves to 12″ long, *two linear bracts* below the umbel, and *many* flowers (but no bulbs). Flowers are pink or lavender, and blooming occurs from *late July to mid-August*. Habitat is dry prairies.

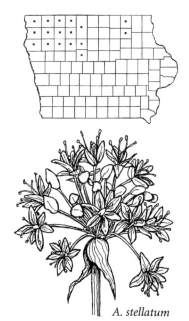

A. stellatum

Yellow stargrass

Star-grass

Hypoxis hirsuta (L.) Cov.

Stem: perennial from a bulb.

Leaves: basal; linear, pointed, with sheathing base; 4″ to 8″ long by 1/4″ wide; lightly long-hairy.

Inflorescence: few flowers at the tip of a hairy branching flower stalk from the base of the plant; the flower stalk two-thirds the length of the leaves.

Flowers: sepals and petals yellow, 3/8″ long, long-hairy, on a tiny, hairy ovary (inferior ovary); six prominent stamens, yellow; flowering from early May to early June.

Fruits: capsule, 1/8″ long, lumpy, green; few black seeds, 1/16″ in diameter; fruiting begins in mid-May.

Habitat: common to frequent on moist to mesic prairies; also in woodland openings and open woods.

Wood lily
Lilium philadelphicum L.

Stem: perennial; 2 1/2′ to 3 1/2′ tall; unbranched; smooth.

Leaves: alternate; linear with tapering base and pointed tip; sessile; 2 1/2″ by 1/4″; smooth above and below.

Inflorescence: umbel of up to four flowers at the stem tip; flower stalks 2″ to 3″ long; flowers held upright; three or four leaves in a whorl just below the inflorescence.

Flowers: petals and sepals orange, 2″ long, with a narrow base and widening above; dark spots on the lower half of petals and sepals; stamens and style raised above petals and sepals; flowering from mid-June to mid-July.

Fruits: capsule; 2″ long by 3/8″ in diameter, cylindrical, with three chambers; fruits mature in late August.

Habitat: infrequent on moist to mesic prairies; also on dry and sandy prairies.

L. michiganense

L. philadelphicum

Michigan lily
Turk's-cap lily
Lilium michiganense Farw.

Lilium michiganense is similar to wood lily, but the leaves are *whorled* (5–7 per node) and smooth above and hairy below. Flowers *face downward on arching stalks* with the petals and sepals *curved backward*. The fruit, a capsule, is *smaller* (1 1/2″ long), *widest above the middle, and tapering to the base*. Flowering is from late June to mid-July. Fruits mature in late August. *L. michiganense* is infrequent on moist prairies and roadsides and in open grassy places to moist, open woods. It is most common in northeastern Iowa.

Bunch-flower

Melanthium virginicum L.
  Veratrum virginicum (L.) Ait. f.

Stem: perennial; 4′ to 5′ tall; branched
  near the top, with long downy hairs
  on the ridges.
Leaves: alternate; linear, with clasping
  bases and sharp tips; 4″ to 18″ by
  3/4″; smooth above and below.
Inflorescence: several racemes from
  the stem tip and upper leaf axils,
  to 10″ long; many flowers on each
  flower stalk; flower stalk with curly
  hairs.
Flowers: petals and sepals white, 1/4″
  long, with narrow base and widening above, two nectar-producing
  glands near the base of each sepal
  and petal; ovary 3/16″ long, with
  three curving style branches; flowering from early to mid-July.
Fruits: capsule, 1/2″ long, with three
  linear lobes; flower parts remain
  attached and turn green; fruiting
  begins in late July.
Habitat: very uncommon on moist
  prairies; also on mesic prairie.

White camass
 Death camass
 Zigadenus elegans Pursh

Stem: perennial; 1 1/2′ to 3′ tall; not
 branched; smooth.
Leaves: alternate, mostly from near the
 base of the stem; linear, tapered to a
 pointed tip; 8″ to 12″ by 1/2″; upper
 stem leaves 2″ to 5″ by 1/2″; smooth
 above and below.
Inflorescence: one to several racemes
 from the stem tip and upper leaf
 axils; each raceme to 8″ long; bract
 at the base of each flower stalk
 about 1″ long.
Flowers: petals and sepals similar,
 white or greenish white, 1/4″ long,
 with two greenish nectaries on the
 upper surface of the sepals and petals; stamens flat against the pistil,
 making a cone-shaped structure
 5/16″ tall in the center of the flower;
 flowering from mid- to late June.
Fruits: capsule, 5/8″ long; oval-shaped,
 tapering toward the tip with divergent style branches; three chambers;
 fruiting begins in early July.
Habitat: very uncommon on moist
 prairies.

A related species, white camass, *Zigadenus glaucus* Nutt., is very uncommon
in northeastern Iowa on *dry, rocky
limestone ridges*. It is very similar in
appearance but with *more branching* in
the inflorescence and with a *shorter
fruit* (about 3/8″ long).

ORCHIDACEAE
Orchid Family

Small white lady's-slipper orchid
Cypripedium candidum Muhl.
ex Willd.

Stem: perennial; 6″ to 12″ tall; lower
stem sheathed with leaves; smooth.

Leaves: alternate; oval, tapering to
both ends, sheathing the stem; 4″
by 1″; smooth above and below.

Inflorescence: single flower at the tip
of the flower stalk, carried above the
leaves; an upright leaf attached just
below the flower; the flower stalk
hairy above.

Flowers: "slipper" (lower petal) white
with purple lines within, 3/4″ long;
upper two petals greenish, linear,
about 1″ long; sepals greenish, upper
sepal 1 1/4″ by 1/4″, lower two sepals
linear, about 1″ long; with inferior
ovary; flowering from early to
late May.

Fruits: capsule, cylindrical, 1″ by 5/16″
in diameter, one chamber; fruiting
begins in mid-June; many! tiny!
seeds.

Habitat: very uncommon on moist
prairies; also known from fens; an
endangered species in Iowa.

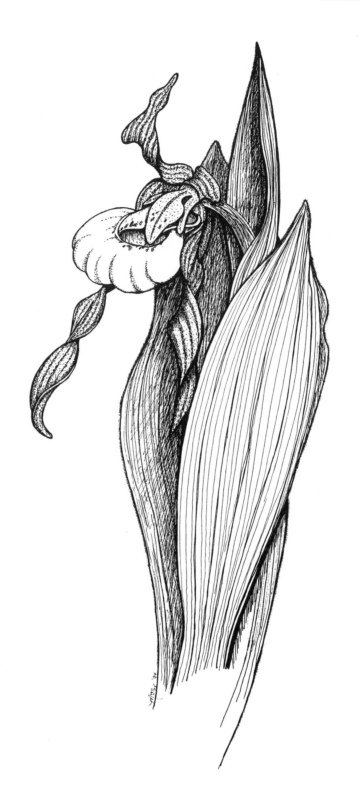

Western prairie fringed orchid

Platanthera praeclara Sheviak & Bowles

Stem: perennial; 2′ to 3′ tall; unbranched; smooth.

Leaves: alternate; oval, tapering to both ends, sheathing the stem at the base; 4″ by 1″, the sheath about 4″ long; smooth above and below.

Inflorescence: spike, about 6″ long; each flower with a 1″ bract below its attachment to the flower stalk.

Flowers: lower petal in three lobes, fringed on the margins, 1″ long, with a spur from the base up to 2″ long; upper petals 1″ long; sepals about 7/16″ long; petals and sepals attached to the top of the linear ovary (inferior ovary), 1″ long; flowering from mid-June to mid-July.

Fruits: capsule, linear, 1″ by 1/8″; flower parts remain attached during early development; fruits develop in August; many! tiny! seeds.

Habitat: very infrequent on moist prairies; apparently not blooming every year; listed as endangered in Iowa and threatened federally.

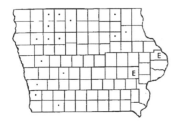

A closely related species, eastern prairie fringed orchid, *Platanthera leucophaea* (Nutt.) Lindley, with smaller flowers, shorter petals, and a shorter spur, is rare in eastern Iowa (designated with E on the map above). It is also listed as endangered in Iowa and threatened federally.

Nodding ladies'-tresses
Spiranthes cernua (L.) L. C. Rich.

Stem: perennial; 8″ to 12″ tall;
unbranched; smooth.

Leaves: basal and on the lower stem;
basal leaves linear, 4″ by 1/4″, tapering to both ends; stem leaves clasping, to 1″ long; smooth above and
below.

Inflorescence: spike, 1 1/2″ to 2″ long,
flowers arranged in tight spirals;
flower stalk hairy.

Flowers: petals and sepals white, about
1/2″ long, attached to the end of the
ovary, 3/16″ long; flowering from
late August to early October.

Fruits: capsule, oval, 5/16″ long by
3/32″ in diameter, one cavity; fruiting begins in mid-September; many!
tiny! seeds.

Habitat: infrequent on moist to wet
prairies.

POACEAE
Grass Family

The grass family has reduced flowers and linear leaves. The flowers (florets) are arranged into spikelets; each spikelet has two bracts (glumes) at its base with one or more florets above. Each floret has two bracts, lemma and palea, which enclose the three stamens and the single pistil that eventually produces the grain. The spikelets are arranged into an inflorescence in two main ways: as a spike with the spikelets crowded on an unbranched axis or as a panicle where spikelets are at the ends of much-branched stalks. Many grasses such as western wheatgrass and Kalm's bromegrass have several florets in each spikelet. Many other grasses such as porcupine grass and switchgrass have a single floret in each spikelet. The leaves alternate across from each other up the stem. The lower part of the leaf is a sheath, which is wrapped around the stem, while the upper portion of the leaf, the blade, diverges from the stem. At the junction of the sheath and the blade is the ligule, a membrane or a fringe of hairs growing from the upper margin of the sheath.

Ligules

*Andropogon
gerardii*

*Panicum
virgatum*

*Bouteloua
curtipendula*

*Schizachyrium
scoparium*

*Elymus
canadensis*

*Sorghastrum
nutans*

Western wheatgrass
Agropyron smithii Rydb.

Stem: perennial; flowering stalk to
2′ tall; from an underground stem;
smooth.

Leaves: sheath smooth; ligule none;
with pointed auricles; blade 8″ by
3/16″, long-tapering to a sharp tip,
with prominent veins, bluish,
smooth.

Inflorescence: spike of alternating
spikelets; 2 1/2″ to 4″ long.

Spikelets: 5/8″ to 3/4″ long; glumes
3/8″ long; several florets, stiff hairs
on the margins of the lemmas, lemmas sometimes short-awned; flowering from mid-June to early July.

Fruits: florets 1/2″ long; spikelets often
fall from the glumes; fruiting begins
in late June; spikelets begin to fall in
mid-July.

Habitat: frequent in dry soils, especially sandy or rocky; on prairies,
roadsides, and open places to open
woods; more common in western
Iowa.

Agropyron smithii closely resembles
quackgrass (*A. repens*), a noxious
weed, which has greener leaves. Making a positive identification is best
done by experts.

Slender wheatgrass
Agropyron trachycaulum (Link) Malte

Agropyron trachycaulum is similar to
western wheatgrass but is *tufted, without underground stems*. The leaves are
somewhat *shorter and more narrow*
(6″ by 3/32″), and the *spikelets are
appressed to the stem, making the inflorescence very narrow*. The spikelets are
often awned to 1″ long. Flowering is
from mid-June to mid-July. Fruiting
begins in late June, but the spikelets do
not fall until September. *A. trachycaulum* is infrequent on dry to mesic prairies to open woods.

A. smithii

A. trachycaulum

Big bluestem
Andropogon gerardii Vitman

Stem: perennial; tufted; flowering stalk
3′ to 7′ tall; smooth.

Leaves: sheath smooth or hairy; ligule
1/4″ tall, thin; blade 1′ to 2′ by 1/4″,
smooth or hairy.

Inflorescence: spikes, three to five at
the tip of the flower stalk, 2″ to 3″
long; resembling the foot of a bird;
turkeyfoot is another common
name.

Spikelets: in twos, the lower is perfect
(male and female), the upper, on a
short, hairy stalk outside the lower,
is male; behind the perfect spikelet
is another hairy stalk that supports
the next pair of spikelets and so forth
to the end of the spike; flowering
from mid-July to mid-September.

Fruits: falling as hairy units which
include the female spikelet and the
stalks in front and behind; 3/8″ long;
fruiting begins in early August; fruits
begin dropping in mid-August.

Habitat: common on moist to mesic
prairies; sometimes in open woods;
also on roadsides and in open
places.

Big bluestem is the most important
grass of the tallgrass prairie.

Side-oats grama

Bouteloua curtipendula (Michx.)
Torrey

Stem: perennial; tufted; flowering stalk
2′ to 3′ tall, lightly hairy.

Leaves: sheath smooth but with stiff,
marginal hairs; ligule 1/32″ long,
fringed; blade with auricles at the
base, stiff marginal hairs near the
base or sometimes only marginal
bumps; 6″ to 8″ long, tapered to
sharp tip.

Inflorescence: many short spikes
attached on one side of the zig-zag
flower stalk; each spike 1/4″ to 3/8″
long; inflorescence 4″ to 8″ long.

Spikelets: 3/16″ long, glumes as long as
the lemmas, one fertile floret per
spikelet, with red anthers; flowering
from mid-June to early July.

Fruits: spikes drop as a unit; fruiting
begins in early July; fruits begin
dropping in mid-August.

Habitat: common on dry prairies and
Loess Hills prairies.

Hairy grama
Bouteloua hirsuta Lag.

Bouteloua hirsuta is similar to side-oats grama but is *shorter* (6″ to 12″ tall). The leaf sheaths are *smooth* with *leaf blades 4″ long* having a few long hairs above and below. There is usually only *one 3/4″ to 1″ long spike per flower stalk.* The flower *stalk extends beyond the florets.* The spikelets are 1/4″ long with the *outer glume with dark, bulb-based hairs.* Flowering is from early to late July. Fruiting begins in mid-July. *B. hirsuta* is infrequent on dry, sandy, and rocky prairies, becoming very infrequent in eastern Iowa.

B. hirsuta

Blue grama
Bouteloua gracilis (Willd. ex HBK.) Lag. ex Steudel

Bouteloua gracilis is similar to side-oats grama but is *shorter* (1′ to 1 1/2′ tall). There are *tufts of hairs at the top of the leaf sheath on each side,* and the blade is 3″ to 10″ long. The *inflorescence consists of one or two one-sided, curved spikes, 3/4″ to 1 1/2″ long.* The spikelets are *purple.* Flowering is from mid-June to mid-July. Fruiting begins in early July, and fruits begin dropping in mid-August. *B. gracilis* is frequent on *dry, gravelly prairies and Loess Hills prairies.*

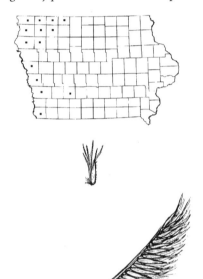

B. gracilis

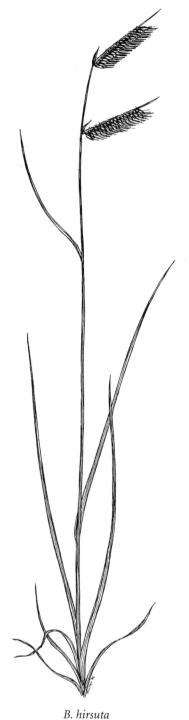

B. hirsuta

Kalm's bromegrass
Bromus kalmii Gray

Stem: perennial; 2 1/2′ to 3′ tall; smooth with hairy joints.

Leaves: sheath closed, hairy, more hairy at the top of the sheath; blade 6″ by 1/4″ and longer, long-hairy below and on the margins.

Inflorescence: narrow, branched, 3″ to 4″ long; with a weak drooping stalk; fifteen to twenty spikelets per inflorescence.

Spikelets: 3/4″ long, ten to twleve florets; glumes and lemmas densely hairy, lemmas awned; flowering from mid- to late June.

Fruits: mature florets falling from the glumes; fruiting begins in early July.

Habitat: very uncommon on upland to dry prairies and in open woods.

Bromus is the only grass genus described in this book that has leaf sheaths where the margins grow together (closed), a characteristic found in the sedges. All other genera have sheath margins that overlap.

Smooth brome, *B. inermis* Leysser, an introduced pasture grass, is very common on roadsides and in other open places. It forms dense sods rather than tufts and is not hairy on the leaves or spikelets.

Bluejoint
Calamagrostis canadensis (Michx.)
Beauv.

Stem: perennial; sod-forming; 3′ to
   5′ tall; smooth.
Leaves: sheath smooth with prominent
   veins; ligule 1/8″ tall, thin; blade
   12″ by 1/4″, long-tapered to a sharp
   point, smooth.
Inflorescence: flower stalk much
   branched, not widely spreading,
   8″ long.
Spikelets: 3/32″ long, glumes purple-
   tipped early; longer than the lemma,
   a tuft of hairs at the base of the
   single floret; flowering from mid- to
   late June.
Fruits: 1/16″ long; fruiting begins in
   late June; fruits begin dropping
   from the glumes in early July.
Habitat: common on moist to wet
   prairies and marshes; often growing
   in dense patches.

Sand reed-grass
Calamovilfa longifolia (Hooker)
Scribner

Stem: perennial; from underground
stems; 3′ to 5′ tall; smooth.

Leaves: sheath smooth, tuft of hairs at
the top; ligule a fringe of hairs, 1/16″
long; blade 8″ to 2′ by 1/2″, smooth.

Inflorescence: flower stalk branched,
not spreading, often partially within
the sheath of the upper leaf, 6″ to
10″ long.

Spikelets: 1/8″ long, upper glume
longer than the lemma, a tuft of
hairs at the base of the single floret;
flowering from late July to mid-
August.

Fruits: 3/32″ long; fruiting begins in
early August; fruits begin dropping
from glumes in mid-August.

Habitat: infrequent on Loess Hills
prairies, also in sandy soils along the
Mississippi River in southeastern
Iowa.

Rosette panic grass
Dichanthelium spp.
  Panicum Subgenus *Dichanthelium*

Species of *Dichanthelium* (all formerly placed within the genus *Panicum*), known as the rosette panic grasses, are very similar in general structure. Several species in this group are found regularly on Iowa prairies. We have chosen to consider four of the most common species (one with two varieties) to illustrate the group. The rosette panicums have proven to be very difficult to separate satisfactorily into discrete species. If you are interested in studying the group further, consult Pohl's "The Grasses of Iowa" or another grass key covering the Midwest.

The rosette panic grasses are short, not over 2′ tall. They bloom in late spring on flower stalks at the tip of the plant. The florets open and shed pollen but seldom set seeds. In early summer additional flower stalks are produced on branches at the middle of the stem. In this second blooming the florets do not open, but they do set seeds without the aid of pollination. During the summer many leaves are produced, but the plant does not grow much taller, resulting in a very bushy plant.

The following guide can be used to identify the four species (one with two varieties) treated here:

If the ligule is longer than 1/8″
  Dichanthelium acuminatum var. *implicatum*
If the ligule is missing or shorter than 1/8″
  and the leaf blade is smooth above and below or with only a few scattered hairs
    Dichanthelium oligosanthes var. *scribnerianum*
  and the leaf blade is hairy only stiff-hairy below
    Dichanthelium linearifolium
  hairy both above and below
    and the spikelets are about 1/16″ long
      Dichanthelium oligosanthes var. *wilcoxianum*
    and the spikelets are about 1/8″ long
      Dichanthelium leibergii

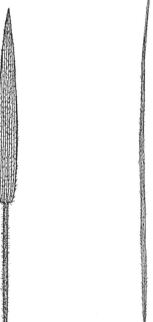

D. acuminatum var. *implicatum*   *D. oligosanthes* var. *scribnerianum*   *D. oligosanthes* var. *wilcoxianum*   *D. leibergii*   *D. linearifolium*

Rosette panic grass
Dichanthelium acuminatum (Sw.)
Gould & Clark var. *implicatum*
(Scribner) Gould & Clark
   Panicum implicatum Scribner

Dichanthelium acuminatum is a tufted
perennial, often with several stems
from the same root crown. The stems
are long-hairy with tufts of hairs at the
nodes. The leaf sheaths are hairy, and
the ligules are of 1/8″ long hairs. The
leaf blade is 2″ to 3″ by 3/8″ and
sparsely hairy above and hairy below.
The inflorescence is up to 3″ long with
purple spikelets 1/16″ long. *D. acuminatum* flowers from early to late June, and
fruiting begins in mid-June. Spikelets
begin dropping in late June. *D. acuminatum* is common on sandy to mesic
prairies, on roadsides, and in upland
woods, becoming very infrequent in
western Iowa.

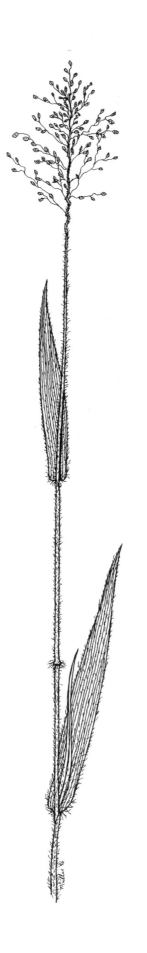

Scribner's panic grass
Dichanthelium oligosanthes (Schultes)
Gould var. *scribnerianum* (Nash)
Gould
  Panicum scribnerianum Nash

Dichanthelium oligosanthes var. *scrib-nerianum* is a tufted perennial often with several stems from one root crown. Leaf sheaths have long, bulb-based hairs or are without hairs. The ligule is tiny. The leaf blades are 2″ to 3″ by 1/4″, smooth above and below, with stiff marginal hairs near the base of the blade. The inflorescence is 3″ long with spikelets 1/8″ long and with short hairs on the glumes. The spikelets have purple bases and stigmas. Flowering is from late May to mid-June. Fruiting begins in early June, and spikelets begin dropping in mid-June. *D. oligosanthes* var. *scribnerianum* is common on mesic to dry prairies, as well as on roadsides and in open woods.

D. oligosanthes
var. *scribnerianum*

D. oligosanthes
var. *wilcoxianum*

Wilcox's panic grass
Dichanthelium oligosanthes (Schultes)
Gould var. *wilcoxianum* (Vasey)
Gould & Clark
  Panicum wilcoxianum Vasey

Dichanthelium oligosanthes var. *wilcox-ianum* is a tufted perennial with a prominent winter rosette. The leaf sheaths have bulb-based hairs. The ligule is of hairs less than 1/16″ long. The hairy leaf blade is 2″ to 4″ by 3/16″. The inflorescence is up to 2 1/2″ long with short-hairy purple (when mature) spikelets 3/32″ long. Flowering is from early to late June. Fruiting begins in mid-June, and spikelets begin dropping in July. *D. oligosanthes* var. *wilcox-ianum* is infrequent on dry, rocky prairies and Loess Hills prairies and is also on sandy prairies, decreasing in frequency to the east and south.

Leiberg's panic grass

Dichanthelium leibergii (Vasey) Freckman

   Panicum leibergii (Vasey) Scribner

Dichanthelium leibergii is a tufted perennial. The leaf sheaths have bulb-based hairs, and the hairy ligule is 1/64″ tall. Leaf blades are 2″ to 4″ by 3/8″ and are hairy above and below. The inflorescence is up to 3″ long, and the hairy spikelets are 1/8″ long. Flowering is from early to late June. Fruiting begins in mid-June, and spikelets begin dropping in late June. *D. leibergii* is frequent on dry upland prairies in northwest Iowa and less frequent to the east and south.

D. leibergii

Rosette panic grass

Dichanthelium linearifolium (Scribner) Gould

   Panicum linearifolium Scribner

Dichanthelium linearifolium is a tufted perennial 8″ to 12″ tall. The leaf sheaths are stiff-hairy and there is no ligule. The leaf blade is 6″ by 1/16″ to 1/8″, narrow and in-rolled near the base and wider above, and stiff-hairy below. The basal leaves are similar to the stem leaves. The inflorescence is 1″ long, and the flower stalks are smooth. Spikelets are 3/32″ to nearly 1/8″ long with prominent veins on the glumes. Flowering is from early to late June. Fruiting begins in mid-June, and spikelets begin dropping in mid-July. *D. linearifolium* is infrequent on dry, sandy, and gravelly prairies in north-central to south-central Iowa and very infrequent elsewhere.

D. linearifolium

Canada wild rye
Elymus canadensis L.

Stem: perennial; tufted; 3′ to 4′ tall;
  smooth.
Leaves: sheath smooth; ligule 1/32″
  long, membranous with tiny hairs
  on the margin; blade 6″ to 8″ by
  3/8″, often somewhat inrolled, at a
  45° angle with stem, auricles
  hooked, smooth.
Inflorescence: spike with two to several spikelets at each node; 4″ to 6″
  long, drooping.
Spikelets: narrow glumes, 1/2″ long,
  tapering into 1/2″ awn, lemma 1/2″
  with 1″ curving awn; flowering from
  mid-July to mid-August.
Fruits: fruiting begins in late July;
  fruits begin dropping in late August.
Habitat: common on upland to lowland prairies; also in open woods,
  on roadsides, and in open places;
  vigorous in disturbed sites.

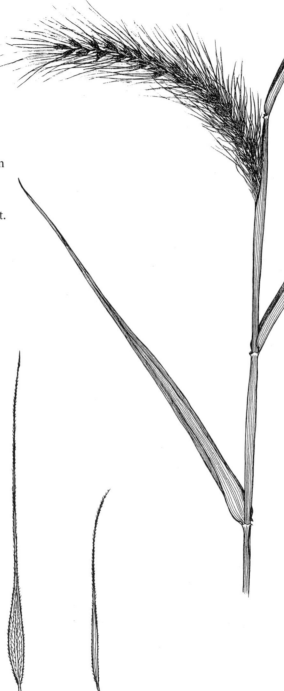

spikelet   glume   lemma

Fescue

Festuca paradoxa Desv.

Stem: perennial; tufted; 2′ to 3′ tall; smooth.

Leaves: sheath smooth; ligule none; blade with oblique base, 8″ by 3/16″, smooth.

Inflorescence: flower stalk branched, spikelets near ends of branches; not spreading, 8″ long.

Spikelets: 3/16″ long, wide and flat; glumes shorter than the lemmas, usually three fertile florets per spikelet; flowering from mid- to late June.

Fruits: fruiting begins in late June; fruits begin dropping from the glumes in August.

Habitat: very uncommon on moist prairies.

June grass

Koeleria macrantha (Ledeb.) Schultes
   K. cristata (L.) Pers. (illegal name)

Stem: perennial; tufted; 1′ to 2′ tall; smooth.

Leaves: sheath smooth or hairy, sometimes marginal hairs at the top of the sheath; ligule 1/32″ long, thin; blade narrow, 4″ to 6″ by 1/16″, hairy above and below.

Inflorescence: branching flower stalk; cylindrical, compact, 2″ to 3″ long, sometimes lobed near the base.

Spikelets: 5/32″ long, the upper (second) glume nearly as long as the florets; usually two florets per spikelet; flowering from early to late June.

Fruits: 1/8″ long; fruiting begins in mid-June; fruits begin dropping in late July.

Habitat: common on sandy and dry prairies, less common on mesic prairies; also on low prairies and in open woods; less common in eastern Iowa.

Marsh muhly
Muhlenbergia racemosa (Michx.) BSP.

Stem: perennial; from underground
stems; 1′ to 2′ tall; smooth or with
tiny hairs.

Leaves: sheath smooth, with a keel
toward the top; ligule 1/32″ long,
thin, with a ragged margin; blade
2″ to 4″ by 1/4″, with prominent
auricles, with an abrupt tip, smooth
or with a few hairs above.

Inflorescence: flower stalk branched,
cylindrical, contracted, lobed; 2″ to
3″ long.

Spikelets: 1/4″ long; glumes awned, the
upper longer than the lower; one
floret; flowering from early August
to mid-September.

Fruits: 1/8″ long; fruiting begins in
mid-August; fruits begin falling
from the glumes in late September.

Habitat: infrequent on dry to moist
prairies; also in open woods and disturbed places.

Plains muhly
Muhlenbergia cuspidata (Torrey) Rydb.

Muhlenbergia cuspidata is similar to
M. racemosa except the leaf blades are
longer (4″ to 6″) and *very narrow* (1/32″
to 1/16″); the inflorescence is *not as
compact, with spikelets nearly sessile on
stalks, 2″ to 4″ long*; spikelets are *smaller*
(1/16″ long); flowering is from *mid-July
to early August*; fruiting begins in early
August; fruits begin falling from the
glumes in September; frequent on *dry
prairies, Loess Hills prairies, and rocky
or sandy prairies*, becoming very infrequent in the eastern part of its range
in Iowa.

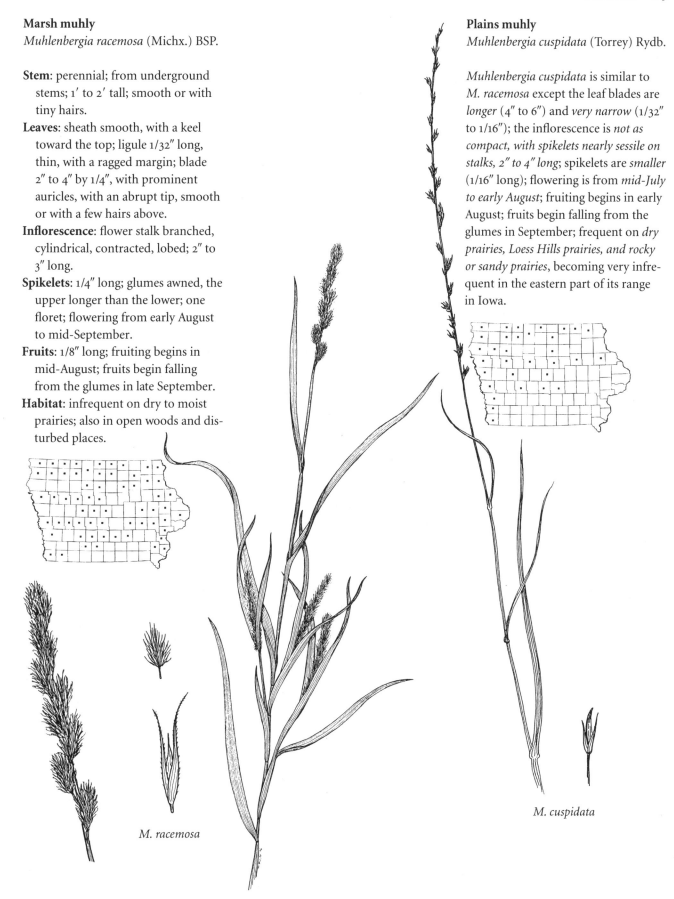

M. racemosa

M. cuspidata

Switchgrass
Panicum virgatum L.

Stem: perennial; from underground stems; 3′ to 5′ tall, smooth.

Leaves: sheath smooth; ligule 5/32″ tall, of dense hairs; blade 12″ by 5/16″, narrow at the base, widening, and then tapering toward the tip to a sharp point, hairy above, especially just above the ligule.

Inflorescence: open, branching flower stalks (panicle); 6″ to 12″ long.

Spikelets: 1/8″ long; upper glume longer than the floret; tips of the glumes, floret, and stigmas purple; flowering from late July to early September.

Fruits: entire spikelet falling as a unit; fruiting begins in mid-August; spikelets begin to fall in early September.

Habitat: common on wet to moist prairies; also on roadsides, stream banks, and other moist, open places.

Switchgrass is commonly included in the seed mixture planted on roadside rights-of-way. Also, it is being planted as a summer pasture and on Conservation Reserve Program land. Genetically selected varieties have been developed for these purposes.

Reed canary grass
Phalaris arundinacea L.

Stem: perennial; from underground stems; 3′ to 5′ tall; smooth.

Leaves: sheath smooth; ligule 1/4″ tall, thin; blade 6″ to 8″ by 1/2″, prominent auricles at the base, smooth.

Inflorescence: branching flower stalks, contracted lobes on short stalks; 4″ to 6″ long; more widely spread during blooming.

Spikelets: 5/32″ long; one floret per spikelet; flowering from late May to mid-June.

Fruits: 1/8″ long; with two tiny, hairy branches at the base of the mature floret; fruiting begins in mid-June; fruits begin dropping from the glumes in late June.

Habitat: frequent in wet, open areas such as stream banks and waterways; sometimes in wet, open woods and occasionally on prairies.

Reed canary grass is sometimes used to stabilize grassed waterways and often escapes into wet roadside ditches and other open, wet places. Improved varieties are much more vigorous than the native form.

Fowl meadow grass
Poa palustris L.

Stem: perennial; tufted; 2′ to 3′ tall; smooth.

Leaves: sheath smooth; ligule 1/16″ long, thin; blade 3″ to 4″ by 1/16″, prominent auricles at the base, tip "boat-shaped."

Inflorescence: wide-spreading, branching flower stalks, linear, to 6″ long.

Spikelets: 3/32″ long, with several florets; upper glume and lower lemmas purple; flowering from late June to mid-July.

Fruits: 3/16″ long; mature floret with a wisp of hairs at the base; fruiting begins in early July; fruits begin falling from the glumes in mid-July.

Habitat: infrequent on moist prairies and marshes; sometimes in moist, open woods; becoming very uncommon in southern Iowa.

Fowl meadow grass is closely related to Kentucky bluegrass, *Poa pratensis* L. (a common lawn grass), but is taller, blooms later, and has a larger inflorescence.

Little bluestem

Schizachyrium scoparium (Michx.)
Nash

    Andropogon scoparius Michx.

Stem: perennial; tufted; flowering stalk
    2′ to 3′ tall; smooth; very flat at the
    base of the stem.

Leaves: sheath smooth or hairy; ligule
    1/16″, fringed; blade 6″ to 8″ by 3/32″,
    long-tapering to a sharp point.

Inflorescence: single spikes from the
    upper leaf axils; about 1″ long.

Spikelets: similar to big bluestem; the
    stalk supporting the male flowers is
    long-hairy, especially toward the tip
    and curved backward; perfect (male
    and female) spikelets with a twisted
    awn, 1/2″ long; flowering from mid-
    July to late August.

Fruits: similar to big bluestem, each
    unit 1/4″ long; fruiting begins in
    early August; fruits begin dropping
    in mid-September.

Habitat: common on dry prairies to
    Loess Hills prairies; less common on
    mesic prairies; often on roadsides
    and in open places.

Indian grass

Sorghastrum nutans (L.) Nash

Stem: perennial; tufted; 3′ to 5′ tall;
smooth.

Leaves: sheath smooth; ligule 1/4″ tall,
stiff, with two points at the tip;
blade about 12″ by 3/8″, narrower
than the sheath at the base, then
widening, with a sharp-tapered tip.

Inflorescence: contracted, branching
flower stalk (panicle), 6″ to 10″ long.

Spikelets: 1/4″ long, hairy stalks on
either side of the floret, floret with
1/2″ twisted awn; lower glume hairy;
flowering from mid-August to mid-
September.

Fruits: yellow-brown; fruiting begins
in late August; mature spikelets
begin falling in late September.

Habitat: common on mesic prairies;
also on dry and moist prairies, on
roadsides, and in open places.

Indian grass is one of the most important grasses of the tallgrass prairie. It is
also one of the more vigorous of the
prairie species in invading stable,
nonprairie areas.

Slough grass

Cord grass

Spartina pectinata Link

Stem: perennial; from strong, underground stems; 4′ to 6′ tall; smooth; emerging shoots very sharp.

Leaves: sheath smooth; ligule 1/16″, fringed; blade to 2′ by 1/2″, arching, smooth, with upward-pointing, sharp, marginal teeth.

Inflorescence: one-sided spikes, 2″ to 6″ long; few to many on each stem; spikelets crowded on the spike.

Spikelets: 1/2″ long, glumes narrow, pointed; single seeded; flowering from mid-July to early September.

Fruits: spikelet falling with the glumes; fruiting begins in early August; spikelets held on the plant into the fall.

Habitat: common on moist prairies; also on wet roadsides and in other open places; often growing in large patches to the exclusion of most other species.

Very few viable seeds are produced by slough grass. Often entire spikes will be barren. In addition, insect larvae often tunnel through the spike, destroying the few seeds that are set.

Dropseed

Tall dropseed, rough dropseed
Sporobolus asper (Michx.) Kunth

Stem: perennial; tufted; 2′ to 3′ tall; smooth.

Leaves: sheath smooth; ligule 1/4″ long, of hairs; blade 6″ to 9″ by 3/16″, tapering to threadlike tip, prominent auricles at the base of the blade, smooth.

Inflorescence: contracted, branching flower stalk (panicle); 4″ to 6″ long; developing within the sheath of the uppermost leaf and becoming more or less exposed with maturity.

Spikelets: 3/16″ long; with a single floret; glumes shorter than the lemma; flowering begins in mid-August.

Fruits: grain 1/16″ in diameter, round; the outer water-absorbent coating very thin; the dark embryo visible in the grain; grains begin falling from the spikelets in late September.

Habitat: frequent in sandy or rocky soils, on prairies, and on roadsides; most common in southeastern and south-central Iowa, very uncommon in northwestern Iowa.

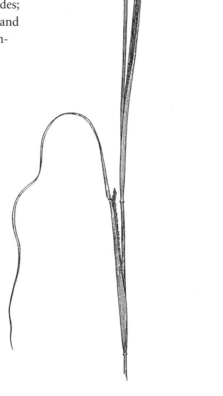

spikelet

glume

Prairie dropseed
 Northern dropseed
 Sporobolus heterolepis (Gray) Gray

Stem: perennial; tufted, older plants
 with a large mound of old stem
 bases; 2′ to 3′ tall; smooth.

Leaves: sheath smooth, tufts of hairs at
 the top on each side at the junction
 with the blade; ligule of hairs; blade
 2′ by 1/16″, becoming very narrow
 near the tip, smooth.

Inflorescence: diffuse, branching
 flower stalk (panicle); to 8″ long;
 very fine and inconspicuous.

Spikelets: 5/32″ long; with a single
 floret; upper glume longer than the
 lemma; flowering from mid-August
 to early September.

Fruits: grain 3/32″ in diameter, round;
 with an outer coating that absorbs
 water and swells; the germ (embryo)
 is dark colored while the endosperm
 is nearly transparent; fruiting begins
 in late August; grains begin falling in
 early September.

Habitat: infrequent on dry to mesic
 prairies; more common in southwestern Iowa.

Older plants have a large mound of old
stems at the center of the plant. The
long leaves droop to the ground all
around, preventing competing plants
from growing close by.

Porcupine grass
Stipa spartea Trin.

Stem: perennial; tufted; 2 1/2′ to 3 1/2′ tall; smooth.

Leaves: sheath smooth; ligule 3/16″ tall, thin; blade 1′ to 2′ by 1/8″, narrower than the sheath at the base, narrowing to a long, threadlike tip, smooth.

Inflorescence: contracted, branching flower stalk (panicle); 4″ to 8″ long.

Spikelets: glumes 1″ long, straw-colored, with long-tapered points; flowering from mid-May to early June.

Fruits: mature grain 1/2″ to 3/8″ long, with a *sharp* point, hairy above the point, and a long awn about 4″ long, the lower 2″ twisted, the outer 2″ bent at an angle; fruiting begins in early June; fruits fall from the glumes between mid-June and early July.

Habitat: common on dry and mesic prairies; also on roadsides and in other open places.

The long awn, which twists as it dries and untwists as it absorbs moisture, catches in plant litter, and the seed is forced into the soil as the awn continues to untwist and elongate.

Green needlegrass
  Feather-bunchgrass
Stipa viridula Trin.

Stipa viridula is similar to porcupine grass, but the leaf blade is *shorter* (4″ to 8″) and *without a ligule* but with tufts of hairs on either side at the top of the sheath. There are *more spikelets in the inflorescence.* The *glumes* are shorter (5/16″), and the mature floret is *much shorter* (3/16″). The awn is also *shorter* (about 1″ long). Fowering is from mid- to late May. Fruiting begins in late May, and mature florets fall from the glumes from late June to mid-July. *S. viridula* is infrequent on dry prairies and railroad rights-of-way in northwest Iowa and very infrequent elsewhere in Iowa.

S. spartea

S. spartea

S. viridula

Iowa Prairies

Open to the Public

NAME	ACRES	OWNERSHIP	COUNTY
*Ames High School Prairie	7	Ames Community Schools	Story
*Anderson Prairie	200	State of Iowa	Emmet
*Cayler Prairie	160	State of Iowa	Dickinson
*Cedar Hills Sand Prairie	90	Nature Conservancy	Black Hawk
Chichaqua Sand Hill Prairie	9	County Conservation Board	Polk
*Clay Prairie	3	University of Northern Iowa	Butler
*Crossman Prairie	10	Nature Conservancy	Howard
*Derald Dinesen Prairie	20	Natural Resource Conservation Service	Shelby
*Doolittle Prairie	25	State of Iowa	Story
Engeldinger Marsh	20	County Conservation Board	Polk
*Five Ridge Prairie	789	County Conservation Board	Plymouth
*Freda Hafner Kettlehole	110	Nature Conservancy	Dickinson
*Gitchie Manitou	144	State of Iowa	Lyon
*Hayden Prairie	240	State of Iowa	Howard
*Hoffman Prairie	36	Nature Conservancy	Cerro Gordo
*Kalsow Prairie	160	State of Iowa	Pocahontas
*Kish-Ke-Kosh	14	State of Iowa	Jasper
*Liska-Stanek Prairie	20	County Conservation Board	Webster
Loess Hills State Forest	20+	State of Iowa	Harrison
Loess Hills State Forest	100+	State of Iowa	Monona
*Manikowski Prairie	40	County Conservation Board	Clinton
*Marietta Sand Prairie	17	County Conservation Board	Marshall
*Mount Talbot	90	State of Iowa	Woodbury
Neal Smith National Wildlife Refuge	8,000	U.S. Fish and Wildlife Service	Jasper
*Nestor Stiles	10	State of Iowa	Cherokee
Ringgold State Wildlife Area	10	State of Iowa	Ringgold
Rochester Cemetery	13	Rochester Township	Cedar
Rockford Fossil and Prairie Park	60	County Conservation Board	Floyd
*Rock Island Preserve	17	County Conservation Board	Linn
*Rolling Thunder Prairie	123	County Conservation Board	Warren
*Sheeder Prairie	25	State of Iowa	Guthrie
Shield Prairie	80	County Conservation Board	Muscatine
*Slinde Mounds	32	State of Iowa	Allamakee
*Steele Prairie	200	State of Iowa	Cherokee
Stinson Prairie	32	County Conservation Board	Kossuth
Sylvan Runkel State Preserve	300	State of Iowa	Monona
*Turin Loess Hills	220	State of Iowa	Monona
Waubonsie State Park	100+	State of Iowa	Fremont
*Williams Prairie	20	Nature Conservancy	Johnson

*Area dedicated as state preserve by the State Preserves Advisory Board.

Glossary

Actinomorphic flowers with radial symmetry with several to many planes through the flower giving mirror images; contrast to *zygomorphic*.

Acute forming an acute angle (less than 90°).

Alternate leaves or branches arranged so that only one arises at the same level (*node*) on the stem (see *opposite* and *whorl*).

Angle ridge (see *Galium* and *Scutellaria*).

Annual having one growth and flowering period, usually germinating in the spring and flowering and dying before fall.

Anther *pollen*-producing part of the *stamen* at the tip of the *filament*.

Appressed lying flat against another structure.

Auricles paired "ears" projecting sideways at the *sheath-blade* junction of grass leaves.

Awn wiry bristle sometimes attached to the tips of *bracts* (*lemmas* or *glumes*) surrounding grass *florets*.

Axil the point where two parts meet, such as leaf and stem.

Basal located near the base of a structure, such as basal leaves located at the base of the stem.

Beak pointed projection on a *fruit* or seed.

Berry fleshy *fruit* with several seed chambers, e.g., tomato.

Biennial having two growth periods and then flowering; germinating and growing the first year, resuming growth the following spring, then flowering and dying that year.

Blade the flattened portion of the leaf.

Bract small leaflike structures, usually associated with *flowers* in the *inflorescence*.

Bulb-based hairs with a swollen base.

Calyx the collective *sepals*, the lower *whorl* of leaflike *bracts* at the base of the *flower*; sometimes colored as in iris and lily.

Capsule a dry *fruit* made up of several sections and opening by splitting from the top.

Catkin an *inflorescence* type where reduced *flowers* (without *sepals* or *petals*) are closely bunched on an unbranched axis.

Chordate heart shaped.

Clasping structure partially surrounding another to which it is attached at the base.

Composite a member of the daisy family (Asteraceae or Compositae).

Compound leaf leaf made up of three or more separate *leaflets*.

Conical cone shaped.

Corolla the collective *petals* of the *flower*.

Corymb flat-topped *inflorescence* whose *flowers* or *heads* are on flower stalks of varying lengths, the outer flowers having longer flower stalks (see *umbel*).

Corymbiform an *inflorescence* in the shape of a *corymb*.

Cyme *inflorescence* type with successive *flowers* arising from below earlier blooming flowers, the oldest flower in the center of the flower cluster.

Disk flowers central *flowers* in *composite* (daisy family) heads, having five-parted, tubular *corollas* rather than straplike *petals* of *ray flowers*.

Ears in grasses, tiny appendages on either side at the base of the leaf *blade* (*auricles*).

Fen boglike area with continuous groundwater seepage, sometimes with hard water, and having a characteristic flora.

Fertile producing functional reproductive structures.

Filament stalk of the *stamen* that bears the *anther*.

Fillaries *bracts* surrounding the heads of *composites* (Asteraceae); also called involucral bracts.

Floral cup a cup at the base of a *flower* with *sepals*, *petals*, and *stamens* attached at the margin.

Floral tube similar to a *floral cup* except the structure is more elongate.

Floret diminutive of *flower*, used to designate grass and sedge flowers.

Flower sexual reproductive structure consisting of *ovary(s)* or *stamens* or usually both and usually with surrounding *petals* and *sepals* producing *fruits* containing seeds.

Follicle dry *fruit* opening on one side; a milkweed *pod* is a follicle.

Frond leaf of a fern.

Fruit structure containing mature seeds; the ripened *ovary*; either dry as in milkweed *pods* or fleshy as in rose *hips*.

Fruiting the process of *fruit* ripening, producing mature seeds.

Glandular areas where hairs or specialized cells secrete substances, such as glandular hairs secreting sticky materials or nectaries secreting nectar.

Glumes pairs of *bracts* attached at the base of grass *spikelets*, lower (first glume) attached just below and across from the upper (second glume).

Habitat type of environment within which an organism lives; examples are moist prairie and open woods.

Head tight cluster of *flowers*, round or elongate; especially in the daisy, pea, and mint families.

Hip fleshy, red growth surrounding the *fruits* ("seeds") of roses.

Hood on milkweed *flowers*, an upward extension of the petal forming a hoodlike structure.

Horn on milkweed *flowers*, a curved petal appendage attached within the *hood*.

Inferior ovary ovary with the *sepals*, *petals*, and *stamens* attached at the top.

Inflorescence collective flower cluster.

Involucre *whorl* of bracts attached below a head of flowers, especially in the daisy family.

Leaf usually flat, green appendage attached to the stem consisting of leaf stalk (*petiole*) and *blade* or *sheath* and blade (in grasses and sedges).

Leaflet segment of a *compound leaf* (a leaf composed of individual *leaflets*).

Lemma larger of two *bracts* that surround the grass *floret* (see *palea*).

Ligule tiny membrane of tissue or row of hairs on the upper side at the junction of the *blade* and *sheath* of grass leaves.

Lobe outward growth, usually rounded.

Mesic soil neither excessively moist nor excessively dry.

Nectar guide colored dots within the *corolla* that apparently assist insects in locating nectar in *flowers*.

Nectary nectar-secreting structures, usually at the base of the petals.

Node portion of the stem from which leaves arise; often somewhat swollen.

Nutlet tiny hard *fruit*.

Opposite situation where two leaves or branches arise on opposite sides of the stem at a *node*, as in members of the mint family (see *alternate* and *whorl*).

Ovary expanded lower portion of the *pistil* where seeds will develop in the *flower*.

Ovate egg shaped; wider or thicker below the middle.

Palea membranous *bract* surrounding the grass *floret*, opposite and smaller than the *lemma*.

Palmate parts originating at a common point and radiating out.

Panicle *inflorescence* type that is highly branched with *flowers* at the tips of the branches.

Perennial producing growth and *flowers* year after year (see *annual* and *biennial*.

Perigynium tubular membrane (open at the top) that encloses the *pistil* in sedges.

Petal white or colored lamina surrounding (or growing from the top of) the *ovary* in the *flower*. Sometimes *sepals* are also petal-like (e.g., lily), and occasionally only one *whorl* is present (*Anemone*), then termed sepals.

Petiole stalk of a leaf, the portion connecting the leaf *blade* with the stem.

Pinnate having parts attached on both sides of a central axis.

Pioneer opportunistic species that quickly invades into disturbed places.

Pistil the female portion of the *flower*; in the center of the flower and containing the *ovary* where the seeds are produced.

Pod generic name for dry *fruits* that open by splitting at one to several places; often used to designate pea family fruits.

Pollen male *spores* produced by the *stamens*.

Raceme unbranched *inflorescence* type with the flowers attached to the axis by short stalks.

Ray straplike "*petals*" surrounding *composite* (daisy family) *heads*, such as the *rays* ("*petals*") on sunflower heads.

Ray flowers *flowers* of composites (daisy family) on the perimeter of *heads* (sunflower) or making up all the flowers of the head (false dandelion); sometimes lacking, as in blazing star (see *disk flowers*).

Receptacle top of the *flower* stalk to which the flower parts are attached.

Recurved bending backward upon itself.

Reflexed curving backward upon itself.

Reticulate netlike.

Rhizome horizontal underground stem, usually giving rise to new shoots.

Rosette cluster of leaves attached to a short stem in a whorled fashion.

Sepal outer (lower) leaflike *bract* at the base of the flower; usually green but occasionally white or colored (see *calyx*).

Sessile attached directly, without a stalk; a *sessile* leaf has no leaf stalk (*petiole*).

Sheath cylinder of tissue surrounding the stem or other plant part, especially prominent on sedges and grasses; in sedges and a few grasses the *sheath* edges grow together forming a cylinder, termed a "closed sheath."

Spike *inflorescence* type with flowers *sessile* on an unbranched axis.

Spikelet cluster of one or more grass *florets*, subtended by two *sterile bracts* (*glumes*); in sedges consisting of *floret* with *perigynium* and single subtending *bract*.

Spindle-shaped widest at the middle and tapering to both ends.

Spore tiny reproductive structure of ferns and horsetails.

Spur elongate, hollow, curved growth, often containing a *nectary*.

Stamen male structure in the *flower* producing *pollen*; almost always more than one and frequently three, four, five, six, or ten to many per flower.

Sterile not producing reproductive structures.

Stigma upper portion of the *pistil* that captures *pollen* (see *pistil*, *ovary*, and *style*).

Stipule paired leaflike *bracts* at the base of the leaf, one on each side of the leaf stalk (*petiole*) at the junction with the stem.

Style stalk at the top of the *ovary* that carries the *stigma* at the upper end.

Tendril hairlike growth that winds around whatever it touches, usually substituting for a leaf or *leaflet*.

Terminal referring to the tip of a structure.

Umbel flat-topped *inflorescence* with *flowers* or flower clusters at the ends of stalks originating at the same point on the stem (see *corymbiform*).

Whorl leaves or branches arranged so that more than two leaves arise at the same level (*node*) on the stem (see *alternate* and *opposite*).

Zygomorphic flowers having bilateral symmetry, that is, having only one plane through the flower, which produces mirror images; contrast to flowers with radial symmetry (*actinomorphic*) with several to many planes giving mirror images.

References

FLORAL STUDIES

Carter, J. 1960. The Flora of Northwestern Iowa. Ph.D. dissertation, University of Iowa, Iowa City.

Carter, J. 1961. A Preliminary Report on the Vascular Flora of Northwestern Iowa. *Proceedings of the Iowa Academy of Science* 68:146–152.

Cooperrider, T. 1962. *The Vascular Plants of Clinton, Jackson and Jones Counties, Iowa*. State University of Iowa Studies in Natural History, Iowa City.

Davidson, R. 1959. *The Vascular Flora of Southeastern Iowa*. State University of Iowa Studies in Natural History, Iowa City.

Eilers, L. J. 1971. *The Vascular Flora of the Iowan Area*. University of Iowa Studies in Natural History, Iowa City, Iowa.

Eilers, L. J., and D. M. Roosa. 1994. *The Vascular Plants of Iowa*. University of Iowa Press, Iowa City. Lists all vascular plants known to occur in Iowa by plant family with common names. Notes on habitat, abundance, and distribution are included.

Fay, M. 1953. The Flora of Southwestern Iowa. Ph.D. dissertation, University of Iowa, Iowa City.

Fay, M. 1953. A Preliminary Report of the Flora of Southwestern Iowa. *Proceedings of the Iowa Academy of Science* 60:119–121.

Gleason, H. A., and A. Cronquist. 1991. *Manual of Vascular Plants of Northeastern United States and Adjacent Canada*. 2d ed. New York Botanical Garden, Bronx, New York. Technical manual of keys and descriptions of vascular plants of the region.

Great Plains Flora Association. 1977. *Atlas of the Flora of the Great Plains*. Iowa State University Press, Ames. Maps showing distribution of the Great Plains flora.

Great Plains Flora Association. 1986. *Flora of the Great Plains*. University Press of Kansas, Lawrence. Keys and descriptions of plants found on the Great Plains.

Hartley, T. 1966. *The Flora of the "Driftless Area."* University of Iowa Studies in Natural History, Iowa City.

Mohlenbrock, R. H. 1975. *Guide to the Vascular Flora of Illinois*. Southern Illinois University Press, Carbondale. Keys and descriptions of Illinois vascular plants.

Mohlenbrock, R. H. 1978. *Distribution of Illinois Vascular Plants*. Southern Illinois University Press, Carbondale. Maps giving the counties where species have been collected in Illinois.

Monson, P. 1959. Spermatophytes of the Des Moines Lobe in Iowa. Ph.D. dissertation, Iowa State University, Ames.

Pohl, R. W. 1967. The Grasses of Iowa. *Iowa State Journal of Science* 40:341–573. Keys and drawings of Iowa's grasses.

Styermark, J. 1963. *Flora of Missouri*. Iowa State University Press, Ames. Keys and drawings of Missouri plants.

Van Bruggen, T. 1959. A Report on the Flora of South-central Iowa. *Proceedings of the Iowa Academy of Science* 66:169–177.

Van Bruggen, T. 1959. The Vascular Flora of South-central Iowa. Ph.D. dissertation, University of Iowa, Iowa City.

Van Bruggen, T. 1976. *The Vascular Flora of South Dakota*, 2d ed. Iowa State University Press, Ames. Keys and descriptions of South Dakota vascular plants.

NATURAL HISTORY

Cooper, T., ed. 1983. *Iowa's Natural Heritage*. Iowa Natural Heritage Foundation, Des Moines, and Iowa Academy of Science, Cedar Falls. Chapters on geology, prairies, forests, wetlands, wildlife, archaeology, and settlement.

Madson, J., and F. Oberle. 1993. *Tallgrass Prairie*. The Nature Conservancy, Arlington, Virginia, and Falcon Press, Helena, Montana. A beautifully illustrated presentation of prairies across North America.

Mutel, C. F. 1989. *Fragile Giants: A Natural History of the Loess Hills*. University of Iowa Press, Iowa City.

Peterson, R. T., and M. McKenny. 1968. *A Field Guide to Wildflowers of Northeastern and Northcentral North America*. Houghton Mifflin, Boston. A classic for identification of wildflowers.

Prior, J. C. 1991. *Landforms of Iowa*. University of Iowa Press, Iowa City. A review of Iowa's landscape.

Pyne, S. 1982. *Fire in America: A Cultural History of Wildland and Rural Fire*. Princeton University Press, Princeton, N.J. Review of the role of fire in forest and prairie habitats.

Runkel, S. T., and D. M. Roosa. 1989. *Wildflowers of the Tallgrass*

Prairie: The Upper Midwest. Iowa State University Press, Ames. Color photographs and descriptions of selected prairie plants.

Shimek, B. 1911. The Prairies. *Bulletin from the Laboratories of Natural History of the State University of Iowa* 11(2):169–240. The first comprehensive review of prairies in Iowa.

Shirley, S. 1994. *Restoring the Tallgrass Prairie: An Illustrated Manual for Iowa and the Upper Midwest.* University of Iowa Press, Iowa City. Guide to species and techniques for prairie restoration.

Thompson, J. R. 1992. *Prairies, Forests, and Wetlands: The Restoration of Natural Landscape Communities in Iowa.* University of Iowa Press, Iowa City. A guide to reconstruction of natural communities.

Weaver, J. E. 1954. *North American Prairie.* Johnson Publishing, Lincoln, Nebraska. After fifty years still an excellent reference on prairies.

Index

Bur Oak Books/Natural History

Birds of an Iowa Dooryard
By Althea R. Sherman

A Country So Full of Game:
The Story of Wildlife in Iowa
By James J. Dinsmore

Fragile Giants:
A Natural History of the Loess Hills
By Cornelia F. Mutel

Gardening in Iowa and
Surrounding Areas
By Veronica Lorson Fowler

An Illustrated Guide
to Iowa Prairie Plants
By Paul Christiansen and Mark Müller

Iowa Birdlife
By Gladys Black

The Iowa Breeding Bird Atlas
By Laura Spess Jackson, Carol A.
Thompson, and James J. Dinsmore

Iowa's Geological Past:
Three Billion Years of Change
By Wayne I. Anderson

Iowa's Minerals:
Their Occurrence, Origins,
Industries, and Lore
By Paul Garvin

Landforms of Iowa
By Jean C. Prior

Land of the Fragile Giants:
Landscapes, Environments, and
Peoples of the Loess Hills
Edited by Cornelia F. Mutel and
Mary Swander

Okoboji Wetlands:
A Lesson in Natural History
By Michael J. Lannoo

Parsnips in the Snow:
Talks with Midwestern Gardeners
By Jane Anne Staw and Mary Swander

Prairies, Forests, and Wetlands:
The Restoration of Natural Landscape
Communities in Iowa
By Janette R. Thompson

Restoring the Tallgrass Prairie:
An Illustrated Manual for Iowa
and the Upper Midwest
By Shirley Shirley

The Vascular Plants of Iowa:
An Annotated Checklist and
Natural History
By Lawrence J. Eilers and
Dean M. Roosa

Weathering Winter:
A Gardener's Daybook
By Carl H. Klaus